Results and Problems in Cell Differentiation

A Series of Topical Volumes in Developmental Biology

1

W0042845

Editors

W. Beermann, Tübingen · J. Reinert, Berlin · H. Ursprung, Baltimore

Managing Editor

H.-W. Hagens, Heidelberg

The Stability
of the Differentiated State

Edited by H. Ursprung, Baltimore

With contributions of

J. Abbott, Philadelphia · A. C. Braun, New York
A. L. Burnett, Cleveland · R. D. Cahn, Seattle · W. Gehring, Zürich
S. D. Hauschka, Seattle · E. D. Hay, Boston · H. Holtzer, Philadelphia
J. W. Lash, Philadelphia · J. R. Whittaker, Philadelphia

With 56 Figures

Springer-Verlag Berlin Heidelberg GmbH 1968

ISBN 978-3-662-34768-3   **ISBN 978-3-662-35089-8 (eBook)**
DOI 10.1007/978-3-662-35089-8

Originally published by Springer-Verlag Berlin Heidelberg New York in 1968.

Softcover reprint of the hardcover 1st edition 1968

Title-No. 3501

Preface

Looking through the tables of contents of recent "Advances" and "Progress" series in Cell and Developmental Biology, one realizes that the majority of these publications still take a systematic, rather than thematic approach for reviewing an area of interest. As a consequence, a given volume in a series often contains a collection of articles that for the specialist appear largely unrelated.

We felt that the time had come to review, periodically, a few of the central issues in Cell and Developmental Biology in highly topical volumes. This seems indicated if a variety of seemingly unrelated avenues of research begin to illuminate one common problem as in the present volume, for example. It is also indicated if a favorite experimental system has been exploited in sufficient depth to warrant a coherent description; some of the volumes planned for the near future fall into this category.

The authors of this volume were instructed not to describe historical background and experimental detail at great length, but to proceed, in each chapter, rapidly to the central issue: "The Stability of the Differentiated State". The book as a whole is meant to demonstrate the bearing of different lines of thought and experimentation on this central topic.

Tübingen, Berlin, Baltimore
January, 1968

W. Beermann, J. Reinert, H. Ursprung
H.-W. Hagens

Contents

Factors Affecting Inheritance and Expression of Differentiation: Some Methods of Analysis

by ROBERT D. CAHN

Dedifferentiation and Metaplasia in Vertebrate and Invertebrate Regeneration

by ELIZABETH D. HAY

The Acquisition, Maintenance, and Lability of the Differentiated State in Hydra

by ALLISON L. BURNETT

The Multipotential Cell and the Tumor Problem

by ARMIN C. BRAUN

The Stability of the Determined State in Cultures of Imaginal Disks in Drosophila

by WALTER GEHRING

Contributors

Dr. JOAN ABBOTT, University of Pennsylvania, Department of Anatomy, School of Medicine, Philadelphia, Pa. 19104 (USA).

Dr. ARMIN C. BRAUN, The Rockefeller University, New York, N. Y. (USA).

Dr. ALLISON L. BURNETT, Development Biology Center, Department of Biology, Western Reserve University, Cleveland, Ohio 44106 (USA).

Dr. ROBERT D. CAHN, Department of Zoology, University of Washington, Seattle, Wash. (USA).

Dr. WALTER GEHRING, Zoologisches Institut der Universität, Künstlergasse 16, CH 8006 Zürich (Schweiz).

Dr. STEPHEN D. HAUSCHKA, Department of Biochemistry, University of Washington, Seattle, Wash. 98105 (USA).

Dr. ELIZABETH D. HAY, Department of Anatomy, Harvard Medical School, Boston 15, Mass. (USA).

Dr. HOWARD HOLTZER, University of Pennsylvania, Department of Anatomy, School of Medicine, Philadelphia, Pa. 19104 (USA).

Dr. JAMES W. LASH, Department of Anatomy, University of Pennsylvania, School of Medicine, Philadelphia, Pa. 19104 (USA).

Dr. J. R. WHITTAKER, The Wistar Institute, Philadelphia, Pa. 19104 (USA).

Oscillations of the Chondrogenic Phenotype *in vitro* *

Howard Holtzer ** and Joan Abbott

Departments of Anatomy and Biological Sciences,
University of Pennsylvania and Columbia University

I. Introduction

To establish perspective, we should like to begin by commenting on "differentiated" *versus* "undifferentiated" cells and their relationship to cell division.

The concept of undifferentiated cells derives from the time when the light microscope was the primary analytical tool of biology. On a molecular or genetic level, however, the distinction between undifferentiated and differentiated cells is less meaningful. Biochemically, "differentiated" cells may be characterized as cells committed to the synthesis of unessential or luxury molecules (Holtzer 1968). All cells produce molecules in common such as tryptophan synthetase, glucose-6-phosphatase, cholesterol, succinic dehydrogenase, the cytochromes or sRNAs: molecules essential for the viability of the cells producing them. Differences between cells are generally based on the production of unessential luxury molecules such as albumin, keratin, hemoglobin, myosin, thyroxin, hyaluronic acid or chondroitin sulfate: molecules which are not essential for the viability of the cells producing them. Unique species of active or inactive mRNAs or rRNAs in egg cells, blastula, or other types of cells would also be luxury molecules if they did not immediately contribute to the survival of the cells. Presumably the succession of precursor cell types observed during embryogenesis is due to the differential production and accumulation of luxury molecules in the newly formed cells. From this viewpoint, precursor cells must be differentiated. The distinction between a terminally differentiated cell and its precursor cell lies in the unique sets of luxury molecules produced by each. No cell is *ever* undifferentiated. All cells must synthesize sets of luxury molecules even if, as yet, they have not been identified.

Consider the myoblast or erythroblast. In the best-studied cases translation for the respective luxury molecules, myosin and hemoglobin, is initiated in the cell at a specific time during the course of one particular cell cycle. Myosin synthesis begins at a particular time in G_1 in a myoblast that has withdrawn from the mitotic cycle (Okazaki and Holtzer 1965, 1966). The withdrawal from the mitotic cycle by erythroblasts which have initiated the translation for hemoglobin is not as abrupt as that of myoblasts (Grasso and Woodard 1966; Marks and Kovach 1966) but follows after an additional one or two mitoses. In the case of the myoblast, the decision to translate for myosin and withdraw from the mitotic cycle is acted on in the same G_1.

* This work was supported by grants from the U.S.P.H.S. and the NSF.
** Career Development Awardee of the U.S.P.H.S.

In the erythroblast, the decision to translate for hemoglobin and withdraw from the mitotic cycle may also be made in the same G_1. However, in the erythroblast, though hemoglobin synthesis is instituted immediately, the decision to withdraw from the mitotic cycle requires another cell division or two to be fully realized. From a mitotic point of view, there are few cells as "dead" as red blood cells. In either case, myoblasts and erythroblasts are the respective descendents of precursor cells which themselves were not synthesizing myosin or hemoglobin. Following a given "quantal" mitosis (HOLTZER 1963, 1968) progeny with a phenotype different from that of the parent cell appears.

Though it is not yet possible in many instances to differentiate microscopically between hematocytoblasts, presumptive myoblasts and presumptive chondroblasts, these progenitor cells are not naive or undifferentiated biochemically. If they were, they would not predictably form only erythroblasts, myoblasts, or chondroblasts respectively, rather than liver, kidney, or skin cells. Replicating hematocytoblasts, presumptive myoblasts, and presumptive chondroblasts are differentiated from all other cells. As such they must synthesize their characteristic luxury molecules – e. g. mRNAs or enzymes which, in their progeny, lead to hemoglobin, myosin or chondroitin sulfate synthesis. How early in embryogenesis cells are channeled into blood-forming, muscle-forming or cartilage-forming lineages is unknown. Hemoglobin-containing cells are present in 30-hour chick embryos and myosin (HOLTZER et al. 1958) is present in heart cells a few hours later. Definitive metachromatic poly-saccharide matrix, characteristic of cartilage, appears much later (HOLTZER 1964; NAMEROFF and HOLTZER 1967). These and many other observations (WILSON 1924) are consistent with the idea that cell lineages are traceable to the zygote and early cleavage stages. Therefore, at any given moment in development all cells are expressing unique genetic programs. The fact that some embryonic or mature cells may be "rechanneled" to express another phenotype demonstrates only that they, *or their progeny*, may be reprogrammed and not that they are, or were, blank, naive, undifferentiated cells.

The succession of transitory phenotypes in a cell lineage must be correlated with, and probably is dependent upon, mitotic activity. Mitosis, both in early embryogenesis and in most renewal systems involving stem cells, leads to a discontinuity of phenotypes. Expression of the terminal phenotype is a function of successive kinds of cell divisions undergone by precursor cells. The asymmetrical partitioning of cytoplasm and cell membranes (WILSON 1924; STOSSBERG 1938; HENKE 1946, HENKE and POHLEY 1952; STEBBINS and SHAH 1960; BEISSON and SONNEBORN 1965) are gross examples of what must occur more subtly at the molecular level. Any number of models could be proposed for the induction of new phenotypes by the asymmetrical distribution of cytoplasmic or cortical factors during mitosis. At this point we simply wish to emphasize that dividing cells are differentiated cells which often yield a progeny with different properties. Cells do not jump from a primitive state to a terminally differentiated state in one step (HOLTZER 1961, 1963, 1964, 1968). Cell speciation is the resultant of mitoses coupled with the response of the nascent daughter cells to a changed environment.

These views lead to the idea that controls effecting genetic regulation of luxury molecules might differ from those effecting the regulation of essential molecules. A developing myogenic cell should survive if it is inhibited from making myosin, but not if inhibited from making its cytochromes. On the other hand, if precursor myoblasts are prevented from dividing, the initiation of myosin synthesis should also be inhibited. We propose that the synthesis of essential molecules is controlled by a homeostatic type of regulatory system allowing a fluctuating production in the numbers of molecules whereas the synthesis of luxury molecules is more qualitative and is governed by an "all or none" type of regulatory system. Although muscle cells under various conditions may synthesize myosin at different rates, *only* committed myogenic cells normally synthesize myosin. Accordingly, we have been trying to learn more of the consequences of uncoupling the phenotypic expression of terminally specialized cells from other aspects of their basic activities.

II. Chondrocytes in Monodisperse Cultures

One of the first questions posed was whether the progeny of chondrocytes were obligated to inherit and express the "differentiated state" of the parent cells in the way in which they were obligated to inherit and use the machinery for synthesizing essential molecules (HOLTZER et al. 1960). Chondrocytes were used because of the ease in obtaining sizeable numbers reasonably free of contaminating cells and because their luxury molecules, chondroitin sulfate and collagen, could be followed histologically and biochemically. Chondrocytes, grown on a plasma clot as monodispersed cells, were periodically challenged to synthesize metachromatic matrix by growing them as pellets *in vitro* or on the chorioallantoic membrane. The monodisperse culture situation discourages matrix formation but encourages cell division. The pellet organ culture provides optimal conditions for chondrogenic expression, but not for cell division.

After proliferating for 2 weeks as monodispersed cells, the progeny of chrondrocytes no longer chondrified when cultured as pellets. In addition, there was minimal $^{35}SO_4$-uptake into chondroitin sulfate or ^{14}C-proline into collagen (PROKOP, PETTINGILL, and HOLTZER 1964; ABBOTT and HOLTZER 1966). Although capable of vigorous multiplication in monodisperse cultures, the large, fragile cells became altered so that chondrogenesis no longer occurred. These cells have been termed "dedifferentiated chondrocytes". As stressed in our original paper, this term is purely descriptive, simply indicating that the progeny did not display the terminal phenotypic properties of their parent cells. Whether the dedifferentiated cells could be induced to chondrify under other conditions was a question we raised (HOLTZER et al. 1961; HOLTZER 1963, 1964), but left for further investigation.

These results suggested that rapid cell divisions might play a role in dedifferentiating the chondrocytes. Additional work (STOCKDALE et al. 1963), however, indicated that factors other than the number of mitoses *per se* operated to alter the phenotypic expression of chondrocytes. Cells multiplying in proximity to chondrifying cells either did not dedifferentiate or did so slowly.

If association with functional chondrocytes induced chondrogenic activity in the progeny of dividing chondrocytes, would proximity to non-chondrogenic cells evoke the same response or would chondrogenesis be suppressed? Freshly liberated chondrocytes, 98—100 per cent of which would chondrify in a pellet in 5 days, were mixed with varying numbers of connective tissue, muscle, kidney, or liver cells. In all cases, proximity of non-cartilage cells inhibited the chondrogenic activity of otherwise

1*

competent chondrocytes (Abbott and Holtzer 1964). Essentially similar results have more recently been reported by Moscona and Garber (1968) for another heterotypic mixture.

Our experiments suggested that foreign substrates or foreign cell membranes interfere with the expression of the chondrogenic phenotype. This led to the possibility that the first step in the alteration of the biosynthetic activities of chondrocytes involves changes in the cell membrane which subsequently could appear as an "inherited" trait.

Changes in the surface of chondrocytes have been correlated with failure to accumulate metachromatic matrix (Abbott and Holtzer 1964, 1966). On a fibrin clot, only chondrocytes retaining or reacquiring their polygonal shape deposit matrix and incorporate large amounts of $^{35}SO_4$. Dispersed chondrocytes rapidly transform into large, stellate, amoeboid cells which do not deposit matrix. After 20 hours on a clot, all dispersed chondrocytes have become fibroblastic. By 46 hours most of the originally postmitotic cells have synthesized DNA and divided. Assuming no volume change, the transformation of cell shape from polygonal to stellate requires, at a minimum, a 4-fold increase in cell surface. Interference with chondrogenic expression, then, is associated with changes in the cell membrane as exemplified by rapid surface expansion, active pseudopodial movements, and enhanced cell division.

III. Chondrocytes in Clonal Cultures

Coon's (1966) finding that chondrocytes produce chondrogenic clones under certain conditions provides another means of analyzing factors permitting or interfering with chondrogenic expression. We have confirmed Coon's observations that in a population of dedifferentiated cells which may not chondrify on a clot in mass culture, there may be a *small* number of cells which can form chondrogenic clones. Of greater interest to us, however, is that a single cell grown in "permissive medium" *invariably* can be forced to yield a progeny, the great majority of which do not chondrify. There is considerable heterogeneity of cell types in clonal cultures that arise from a single chondrocyte. The following experiments illustrate the many problems in interpreting the behavior of cells grown under "clonal" conditions.

Each of several Cooper dishes was seeded with a *single* chondrocyte. The progeny from such a single chondrocyte provided a cell line which was replated at clonal densities (10^3 cells/dish) as described in Abbott and Holtzer (1968) and Chacko and Holtzer (1968). Irrespective of the age of the cultures, the variation in numbers of cells per clone was great. While approximately 80 per cent of the cells survived for more than a week, less than 40 per cent gave rise to clones with appreciable numbers of cells. After 14 days, many were still isolated cells; the remainder were in clones whose cell numbers ranged from two to many thousands. This variability contrasts with the uniformity of plaque size produced by a homogeneous population of microbes.

A spectrum of cell "types" appears in all clonal dishes (Figs. 1—3) their relative numbers varying from clone to clone. In the following discussion all types other than polygonal (the functional chondrocyte) are referred to as stellate, fibroblastic, altered, or dedifferentiated cells. The polygonals replicate more rapidly than stellates for the first two weeks of culture. Thereafter they tend to replicate more slowly. As the stellate, or altered cells, continue to replicate, eventually they dominate the population.

Scoring clones as chondrogenic or non-chrondrogenic may be misleading. In some clones, less than 5 per cent of the cells are surrounded by matrix, in others 95 per cent are associated with matrix. After 14 days, a clonal plate may have 70 percent polygonal cells and only 30 per cent stellate cells. After 4 weeks in the same dish, the relative number of polygonals may dwindle to 10 per cent of the total population, 90 per cent now being stellate. Clearly chondrogenic cells would be selected for if plates were sacrificed at the end of two weeks, whereas equivalent dishes sacrificed at the end of the month would select for stellate cells.

The emergence of dedifferentiated cells in a clonal dish, all of whose cells were derived originally from a *single cell*, unequivocally demonstrates that functional chondrocytes may produce at least two kinds of progeny: one that is polygonal, adherent to its neighbor, and associated with the production of metachromatic matrix, and one that is fibroblastic, amoeboid, and not associated with matrix. It is to be stressed that this shift in phenotypic expression occurs in an environment capable of supporting chondrogenesis (i.e. in a "permissive" medium).

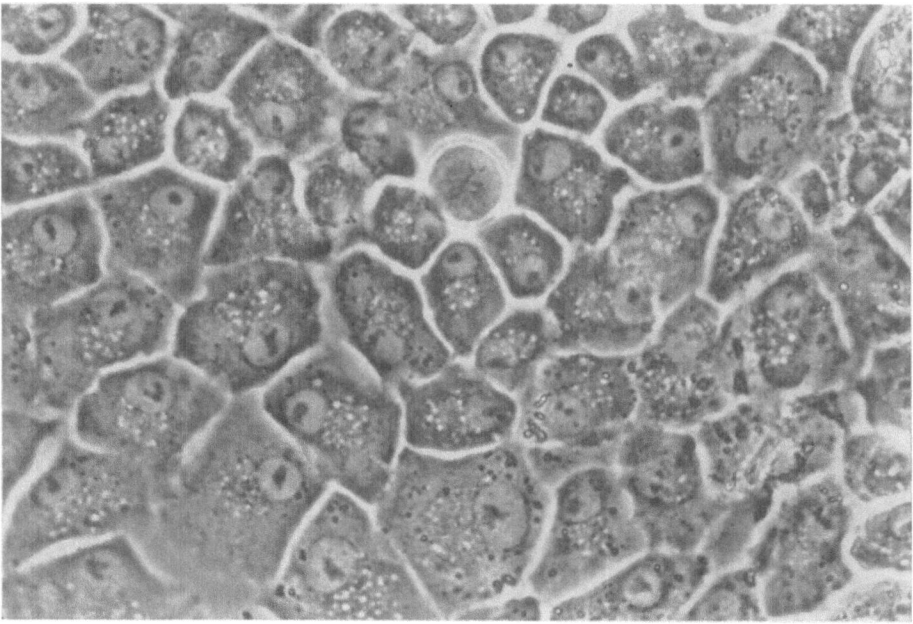

Fig. 1

Figs. 1—3. Phase photomicrographs of living cells from three different areas of a 3-week old clone. All cells were derived from a single chondrocyte. Observe in Fig. 1 the epithelioid nature of the cells in the central zone of the clone. These polygonal cells are often associated with metachromatic matrix. Figs. 2 and 3 depict the variation often observed on the periphery of a clone; here the cells are stellate. Note the expanded surface, conspicuous pseudopodia and microprojections of the stellate or altered cells. Observe the rounded cell in mitosis in the center of the clone in Fig. 1; if fixed and stained this cell in mitosis would be surrounded by metachromatic matrix. These photographs were all taken at the same magnification.

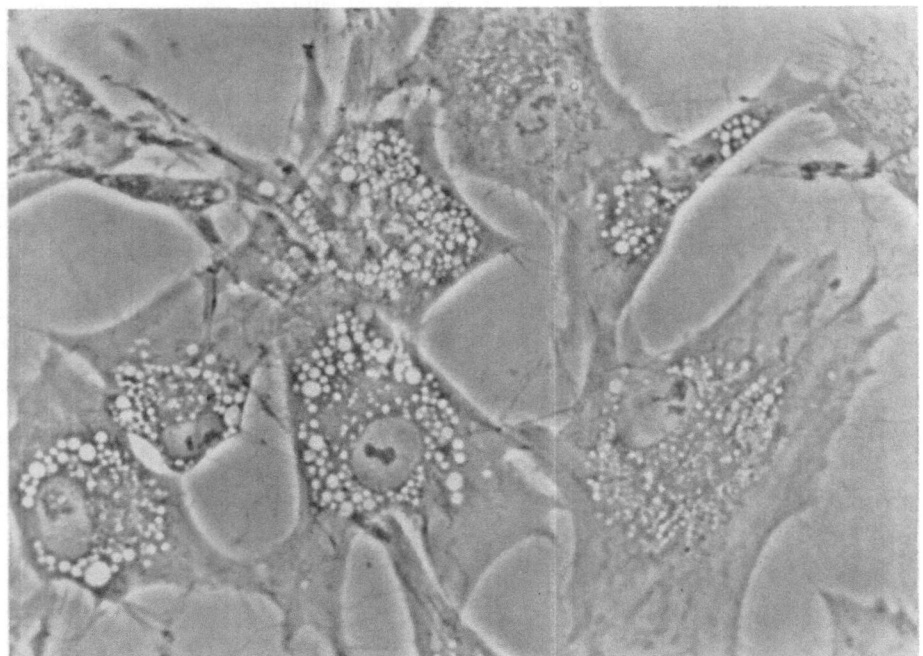

Fig. 2

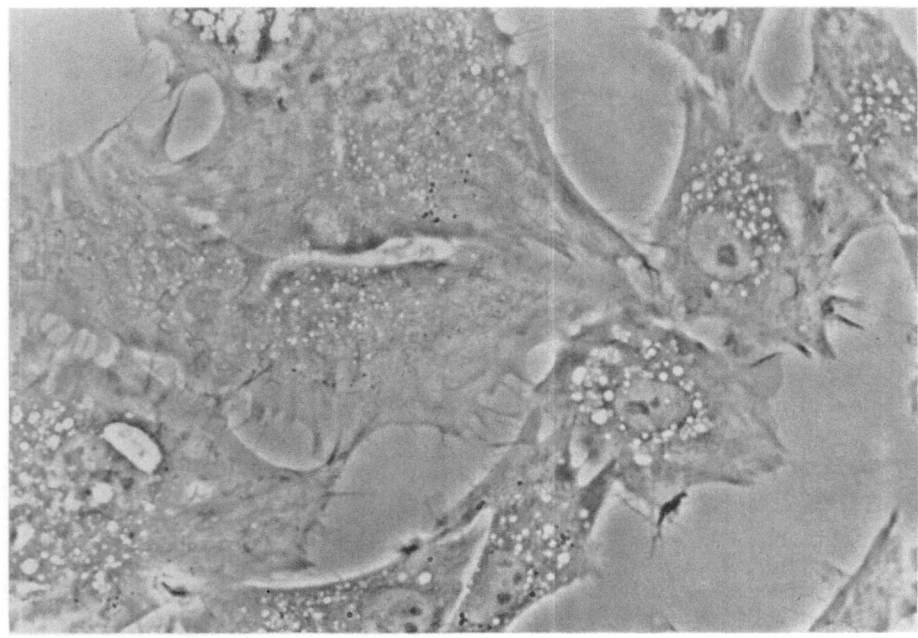

Fig. 3

Do all or even most of the altered chondrocytes revert to polygonals when re-cultered under clonal conditions? The answer to this depends on what happens in clonal cultures and how "plating efficiency" is monitored and defined. In primary clonal plates, polygonal chondrocytes readily detach from the substrate, float around

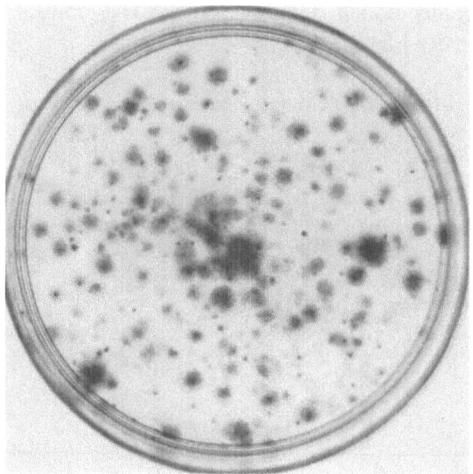

Fig. 4. A single freshly liberated chondrocyte was placed in F-10 in a 60 mm plastic dish. This is a photograph of a toluidine blue stained dish after 33 days of growth. Many of these clones display metachromatic matrix

and reattach to form new clones (Table 1). As a consequence, counting clones yields little information about the behavior of each of the original cells in the population. Figure 4 shows a 60 mm petri dish into which one freshly liberated chondrocyte had been placed 33 days previously. There are over 200 chondrogenic clones spotted over the dish. Obviously, it is absurd to describe the "plating efficiency" of a single cell as 20,000 per cent.

Table 1. *Proliferation of clones from 1—2 cells/plate. 1—2 freshly liberated cartilage cells were plated in 60 mm plastic petri dishes. The number of clones/plate were counted after various lengths of time in culture*

Series	Number of cells plated	Age of culture (days)	Number of clones (>25 cells)
B_4	1	33	228
	1	33	3
	1	39	81
B_2	1	32	28
	1	46	4
	1	46	1
B_5	1	8	3
	1	47	250
B_1	2	35	148
	3	45	531

If one cell can give rise to such a large population of chondrogenic clones, only a few such "sleeper" cells need remain in a dedifferentiated population to give rise to a new chondrogenic population. Under these conditions determination of "plating efficiency" is misleading. The number of clones formed does not represent the number of cells in the original inoculum which were able to form clones. Similarly, the number of chondrogenic clones does not reflect the number of chondrogenically competent cells originally plated.

Table 2. *Behavior of single cell cultures. Single cells were obtained from originally chondrogenic cell strains which had been subcultured 1—4 times. Each plate was inoculated with one cell and scored for presence of cells after 2, 7 and 14 days. The chondrogenic phenotype was scored as positive if any polygonal cells or metachromatic cells were present*

	Number of plates with cells				Number of plates with chondrogenic cells
	Day: 0	2	7	14	
Series	Cell number: 1	1—2	>1	>25	
$B_1 6SC_1$	10	7	6	3	2
$B_1 5SC_1$	8	4	3	1	0
$B_1 6SC_1 7SC$	10	9	5	0	0
$B_1 6SC_2$	5	5	5	3	3
$B_1 6SC_4$	9	9	6	3	1
Totals	42	34	25	10	6

Another approach to the problem of the reversibility of altered cells was to plate single cells from established cultures, many of whose cells appeared dedifferentiated. After 48 hours 81 per cent of the single cells adhered and many had begun to divide. Of these only 20 per cent gave rise to clones of 25 cells or more (range 25 to several thousand cells after 3 weeks) and, of these, approximately 50 per cent were polygonals. Clearly single, altered cells from clonal plates do not clone well after isolation. The observation (Coon 1966) that when dedifferentiated cells are grown under clonal conditions they *all* revert to functional chondrocytes rests on "plating efficiency" experiments and is not in accord with our observations. The emergence of the dedifferentiated cell is not a simple function of the numbers of mitoses the cells have undergone (Stockdale et al. 1963; Holtzer 1964). Dedifferentiated cells may appear in goodly numbers after only 10 days *in vitro*.

Changes in the cell membranes of altered cells are reflected not only in their flattened shapes, but also in their adhesivity to the plastic. Whereas polygonal cells are easily detached from the substrate by trypsinization, the altered cells adhere tenaciously, requiring a longer digestion time to be released.

There is evidence from other experiments that the conditions which favor growth of chondrogenic clones select against the growth and expression of other types of cells. The somites of 4-day chick embryos consist of myogenic cells, connective tissue cells, and chondrogenic cells whose relative population sizes are unknown (Holtzer 1968). When suspended somites were cloned on plastic, the majority of clones that emerged after 2 weeks were chondrogenic (Matheson and Holtzer, unpublished results). The remaining cells were fibroblastic. Whether they represent connective tissue cells, repressed myogenic cells, or altered chondrogenic precursor cells has not been determined. Neither the myogenic nor connective tissue cells formed as many clones nor multiplied as vigorously as did the chondrogenic cells. Clearly many of the myogenic and connective tissue cells fail to survive under these particular clonal conditions. In this respect, the behavior of dedifferentiated cells does not differ from that of non-chondrogenic cells grown under these clonal conditions.

Altered chondrocytes which do not grow well in single cell cultures do grow fairly well at higher densities, albeit more slowly than do the polygonals. In high density plates, clusters of dividing polygonals may appear after 8 or 9 days. The localized nature of these emerging polygonals suggests they are derived from one or two cells. The majority of altered chondrocytes continue to produce altered cells.

The initial behavior of chondrocytes plated at clonal density on a fibrin clot is different from that on plastic. On a clot, daughter cells do not associate in a compact clone, but migrate away from each other. If they continue to multiply, the area of contact between adjacent fibroblastic cells increases and eventually many cells revert to polygonals and deposit matrix. On plastic, chondrocytes generally remain polygonal and non-motile. After division, daughter cells associate to form tight epithelioid mosaic clones (Fig. 1). This epithelioid layer morphologically and functionally appears similar to a perichondrium. Though not proven, we suspect the perichondrial-like arrangement is crucial for the expression of the chondrogenic phenotype and for the rapid multiplication of the polygonal cells under these conditions. This epithelioid arrangement might keep the cells functioning as perichondrial stem cells (ABBOTT and HOLTZER 1966 b).

According to this interpretation, there is a mutual inductive interaction (HOLTZER 1963, 1964) whereby cells in the perichondrial-like sheet constantly reinforce nascent daughter cells to behave as chondrocytes. Cells which lose their contacts with polygonal cells for modest periods are readily induced to revert to polygonals. Kept isolated for longer periods, they undergo changes, marked by changes in their adhesiveness to one another, so that they do not revert to the functional state when again making contact with polygonals. Probably the high molecular weight fraction from embryo extract which COON (1966) and COON and CAHN (1966) claim render a medium "non-permissive" prevents the cells from adhering to each other to form the perichondrial-like colony.

There are additional observations that make it unlikely that the "H-factor" specifically blocks the synthesis of chondroitin sulfate (COON and MARZULLO 1967). Maximal ratios of chondroitin sulfate/DNA are obtained in pellets without cell division in the presence of embryo extract. Dividing chondrocytes grown at high density on a clot in Leighton tubes in the presence of embryo extract also form masses of cartilage. The inhibition by the H-factor is only observed when *small* numbers of cells are grown on *plastic*. This alone makes it unlikely that the postulated high molecular weight molecule specifically blocks an intermediate step in the synthesis of chondroitin sulfate. There is additional evidence that chondrocytes undergo changes *in vitro* in permissive medium. For example, freshly liberated chondrocytes synthesize approximately 4 times the amount of chondroitin sulfate/DNA than do the second generation of sub-cultured cells grown in the same medium. With increasing length of time *in vitro*, though the cells retain their functional properties, there is a diminution in the production of polysaccharide per cell. This diminution is not directly related to the numbers of generations produced by the cells.

Matrix-producing cells in epithelioid sheets divide (Fig. 1). It has been shown that cells in S, G_2, or M do not synthesize myosin (OKAZAKI and HOLTZER 1966; BISCHOFF and HOLTZER 1968). Question: Do chondrocytes synthesize chondroitin sulfate in S, G_2, or M? Experiments using the obvious combinations of ^{3}H-thymidine, $^{35}SO_4$, and colchicine have demonstrated that cells in M do not incorporate $^{35}SO_4$. This is of considerable interest, for presumably the sulfating systems and enzymes are present (D'ABRAMO and LIPMANN 1957). In addition, chondrocytes in M do not incorporate ^{3}H-leucine, ^{3}H-uridine, or ^{14}C-glucose into large, acid-insoluble molecules (ABBOTT and HOLTZER, unpublished observations).

The fact that polygonal cells in S and G_2 incorporate $^{35}SO_4$ (ABBOTT and HOLTZER, unpublished observations; CAHN and LASHER 1967) is not easily interpreted. The poor resolving properties of labelled sulfate require very critical autoradiographic techniques to allow the unequivocal conclusion that the labelled matrix surrounding a given cell was, in fact, secreted by that cell. Very thin sections are essential to demonstrate whether the $^{35}SO_4$ is in, or around, a cell. The labelled matrix could be deposited by contiguous cells not in S or G_2. Labelled sulfate is incorporated into the cell and secreted into the matrix within 20 minutes. Consequently, exposure to thymidine and sulfate, to determine concurrent incorporation, must be delivered as a very brief pulse. Lastly, these experiments may reflect only the incorporation of sulfate which is common to all cells and which may or may not indicate the synthesis of chondroitin sulfate (NAMEROFF and HOLTZER 1967; HOLTZER 1968).

In summary, there is no question that functional chondrocytes are *selected* for when grown under clonal conditions. As part of a perichondrial-like system, they proliferate and express their chondrogenic properties longer than if grown on a clot as monodisperse cells. However, it is equally clear that a single chondrocyte may give rise to generations of cells which need not display the properties of chondrocytes in the situations thus far tested. The dedifferentiated chondrocytes appear and multiply in permissive media. Since dedifferentiated and functional cells, all derived from the same parental cell, grow side by side in the same dish, clearly dedifferentiation is not due uniquely to the suppressive action of some exogenous, large-molecular weight fraction in embryo extract. Cloned cells are not phenotypically equivalent. Judging from the variegated morphologies of the dedifferentiated cells, it is conceivable that there are several classes of altered chondrocytes.

IV. Behavior of Chondrocytes in 5-Bromodeoxyuridine (BUDR)

5-Bromodeoxyuridine (BUDR) inhibits the synthesis of matrix by cloned chondrocytes. If BUDR is added to established chondrogenic clones, the polygonal cells transform into altered stellate cells within 4 to 5 days. The incorporation of ^{14}C-glucose or $^{35}SO_4$ into chondroitin sulfate rapidly diminishes in treated cultures. The cells multiply in BUDR but are flatter than control cells (Fig. 5). They do not adhere to one another, but do cling to the substrate (ABBOTT and HOLTZER 1968).

It has been suggested that BUDR interferes with myogenesis by acting on the cell membrane (STOCKDALE et al. 1963; OKAZAKI and HOLTZER 1965). Figure 6 and 7 and Table 3 provide evidence that BUDR does affect the cell membranes of chondrocytes. Chondrocytes in F-10 do not adhere to the plastic of Falcon's non-tissue culture dishes. When BUDR is added, most of the cells attach by day 5 or 6. The surface area of many BUDR-treated cells is anywhere from 4 to 20 times that of the polygonal controls. BUDR induces giant cell formation among myogenic cells as well.

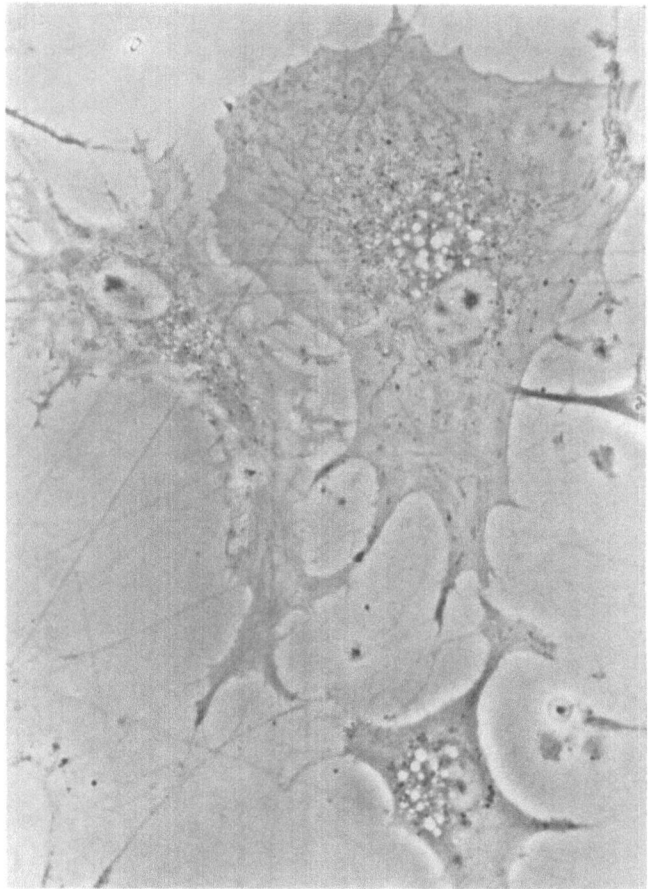

Fig. 5. A photomicrograph of 3 living cells in a culture exposed to BUDR for 6 days. All the cells were originally derived from one chondrocyte. The magnification of this photomicrograph is similar to that of Figs. 1—3; note the enlarged surface area. There is no readily detectable difference between these cells and those altered or dedifferentiated in normal medium in older clones

There is evidence that after one round of DNA synthesis in BUDR myogenic cells are unable to fuse or synthesize myosin and actin (BISCHOFF and HOLTZER, unpublished observations). After a period in BUDR equivalent to one complete mitotic cycle, chondrocytes with one strand of BU-DNA probably cease to synthesize normal amounts of chondroitin sulfate.

The effect of BUDR on cell replication is still unclear. For the first 4 or 5 days, multiplication in BUDR is similar to that in the control population. Thereafter, the rate of multiplication of the BUDR-treated cells lags behind that of controls. The large, flat BUDR-altered cells often appear unable to retract their cell surfaces in preparation for mitosis. Asymmetrical division of cytoplasm and cell surface is very common. Often portions of the cell surface adhere so tenaciously to the plastic substrate that parts of the cells are amputated as the cell migrates or goes through mitosis. Tripolar metaphase plates are observed. Further work will be required to determine how these mechanical factors affect the rate of multiplication.

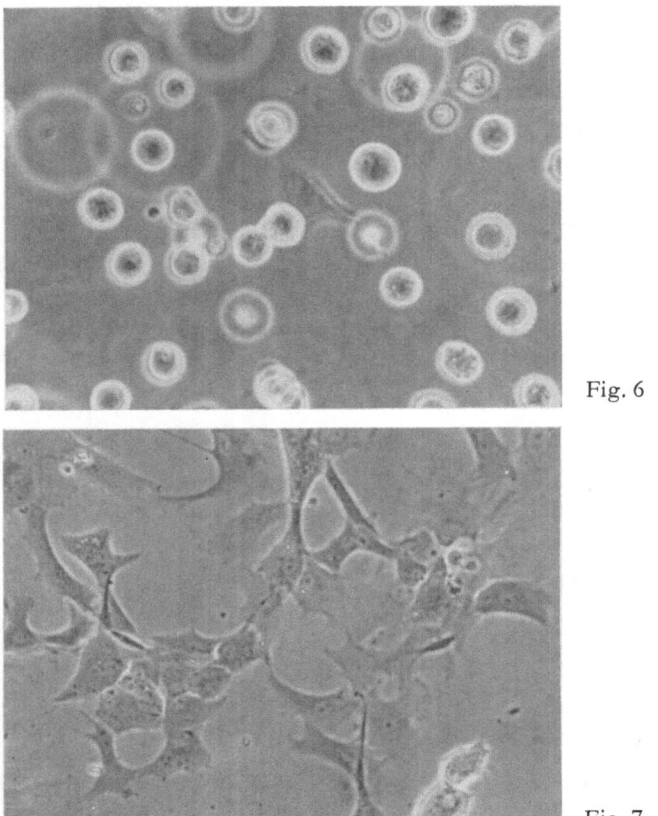

Fig. 6

Fig. 7

Figs. 6 and 7. Subcultured polygonal cells all derived from a single cell, were suspended and 6×10^3 cells introduced into 30 mm nontissue culture dishes. Fig. 6 is the control dish after 9 days of growth. The untreated cells float about but rarely attach to the substrate. Almost all cells in BUDR have attached

Preliminary experiments suggest that 5-bromodeoxycytidine alters chondrocytes in the same manner as 5-bromodeoxyuridine. Although incorporated into DNA, there is no direct evidence at present that the primary effects of these analogues are due to interference at the level of gene transcription (ABBOTT and HOLTZER 1968).

Do all or most chondrocytes, dedifferentiated in BUDR, revert to normal chondrocytes when removed to normal medium? After 10 days in BUDR, altered cells were replated in fresh medium. Following several days of multiplication, many, but not all, displayed the chondrogenic phenotype. Similar results have been observed for muscle (OKAZAKI and HOLTZER 1966; COLEMAN and COLEMAN 1966).

Table 3. *Effect of BUDR on cell adhesivity. 6 × 10³ chondrocytes were plated in non-tissue culture plastic petri dishes. Experimental cultures were exposed to 50 μg BUDR/1 ml from 1—9 days. The number of attached and floating cells were counted separately and the average taken from 2 dishes each of experimental and control plates*

| | Attached cells | | Floating cells | | Total cell |
	Number	Percent	Number	Percent	Number
Controls	4×10^3	8	47×10^3	92	51×10^3
BUDR	100×10^3	88	14×10^3	12	114×10^3

V. Suppression of Chondrogenesis in Mixtures of Functional Chondrocytes with Altered Chondrocytes

Our working hypothesis is that the synthesis of chondroitin sulfate is gradually repressed in chondrocytes that are not immobilized by contiguous functional chondrocytes. Dedifferentiated chondrocytes display surface properties which readily differentiate them from progenitor cells. They do not clone and they do not aggregate with one another. Functional chondrocytes, spun down into a pellet, adhere to one another and within 24 hours synthesize goodly amounts of chondroitin sulfate. Pellets of dedifferentiated cells are much less cohesive and fail to synthesize goodly amounts of chondroitin sulfate even if cultured for a week.

Experiments (ABBOTT and HOLTZER 1964) involving the mixing in monolayer cultures of functional chondrocytes with kidney, liver, or connective tissue cells demonstrated interference with matrix deposition by the proximity of non-chondrogenic cells. Question: Would dedifferentiated or BUDR-altered chondrocytes mixed with functional chondrocytes also exhibit interference. The following is a summary of unpublished experiments by CHACKO and HOLTZER directed to this question. Mixtures of 0.5×10^6 polygonal chondrocytes and 1×10^6 altered cells grown in petri dishes do not display polygonal cells after the first or second day of culture. An unexpected result of mixing polygonal chondrocytes with BUDR-altered cells was the emergence of spindle-shaped cells. For the duration of these experiments the parental cultures remained respectively polygonal and altered. When recultured at high density, the mixed cultures remain predominently spindle-shaped. They behave this way in 3rd generation cultures as well. However, when aliquots of the same mixed cultures are plated at clonal densities (10^3 cells/dish), epithelioid clones as well as large numbers of fibroblastic cells appear. Mixtures grown as pellets do not display functional chondrocytes. However, if the ratio of altered cells is changed from 1:2 to 3:1 patches of chondrocytes appear.

In confrontation experiments dedifferentiated chondrocytes present to functional chondrocytes a micro-environment as foreign as that of liver, kidney, or connective tissue cells. Whether this involves cell surfaces of interacting cells or micro-exsudates, is, of course, unknown. These mixture experiments demonstrate that a *small* number of cells which do not chondrify for varying periods, may do so when allowed to form a clone.

VI. Conclusion

It has been stressed (HOLTZER 1961, 1963, 1964) that dedifferentiated chondrocytes need not represent a reversion to a more primitive, embryonic cell. An altered chondrocyte must possess some unique metabolic programs different from those of other cells, but not necessarily the same as its precursor cells. Our experiments have demonstrated that chondrocytes may oscillate between functional dividing chondrocytes and repressed or altered dividing chondrocytes. However, under certain conditions, altered chondrocytes do not readily revert to polygonal matrix producers. The critical step affecting this loss of reversibility is as yet undefined. Only further work will reveal the basis for the retention or loss of the "chondrogenic memory" by chondrocytes and their progeny. Obviously, chondrogenesis could be blocked at the level of transcription, translation, or polysaccharide deposition. In any case, it would be worth knowing the molecular nature of the block or blocks induced by growth on fibrin clots, in "H-factor", or in BUDR.

Is the kind of inhibition seen *in vitro* a matter only of academic interest and not really germane to *in vivo* processes? We think not. If the transformation of functional chondrocyte to altered chondrocyte is not a "true" differentiation, how operationally, is it to be distinguished from what occurs in other progenitor-descendent relationships obtaining during embryogenesis?

Depending upon exogenous conditions, chondrocytes multiply with or without expressing their chondrogenic genome. This formulation confronts us with the ambiguous literature on embryonic induction. The chorda mesoderm induces the overlying epidermis to express its capacity to differentiate into nerve cells. By the same criteria, the spinal cord or notochord induce somite cells to chondrify (HOLTZER and DETWILER 1952). It has been suggested (HOLTZER 1961, 1963, 1964, 1968) that inducing tissues do not instruct, but only permit, responding cells to synthesize particular luxury molecules. Is the induction of chondrogenesis in altered cells by the removal of an H-like factor, related to the induction of chondrogenesis by spinal cord or notocord? Precursor chondrogenic cells, which do not synthesize chondroitin sulfate may be present from very early embryonic stages and the inducing tissues may simply serve to stimulate their rapid multiplication. Claims that specific cartilage inducing factors have been isolated from spinal cord or notochord (LASH et al. 1962; ZILLIKEN 1967) or that there is such a phenomenon as the formation of "spontaneous cartilage" in a population of somites (LASH 1967) should be viewed with these considerations in mind (HOLTZER 1964, 1968).

This double choice situation – division with or without expression of terminal phenotype – is clearly related to the behavior of hemopoietic stem cells in bone marrow (LAJTHA 1966) or in the seeding of an X-irradiated spleen (McCULLOCH, TILL, and SIMINOVITCH 1966).

Lastly, the conditions leading to the dedifferentiated or altered chondrocyte may be similar to those leading to senescence. Though rapidly dividing, the cells in the epithelioid sheet may be protected from the effects of aging, whereas the isolated cells are not similarly spared.

These experiments originally were initiated to determine whether, in a terminally differentiated cell, the synthesis of luxury molecules could be dissociated from the synthesis of essential molecules. If so, what would be the properties of such altered cells? Though much is yet to be done with this type of *in vitro* analysis, the significance of minor micro-environmental factors in permitting cells to express different aspects of their genetic repertoire is clearly indicated. Devising culture conditions which preserve the production of luxury molecules is very useful. But it will be even more instructive to study the mechanisms whereby minor changes in culture conditions switch off the production of luxury molecules *without killing the cells*. Much information has accumulated on the intracellular machinery concerned with biosynthesis. In the near future comparable efforts must be made to learn how exogenous cues trigger the appropriate responding control system.

Devising culture conditions that permit liver cells to synthesize albumin, cartilage cells to synthesize chondroitin sulfate, or muscle cells to synthesize myosin, undoubtedly will lead to interesting informations about how specialized cells synthesize their respective terminal luxury molecules. These are basic problems in biochemistry. They are not, however, the basic problems of cell differentiation. The major problems in cell differentiation are over when cells synthesize their first molecules of albumin, chondroitin sulfate, or myosin. The major problems in differentiation involve the sequential activation of regulatory complexes preparing precursor cells to undergo quantal mitosis to produce cells which for the first time synthesize their luxury molecules. The appropriate cell to study is not the cell making chondroitin sulfate or myosin, but rather those cells in the lineage of transitory phenotypes leading to the terminal cell type. Operationally the dedifferentiated chondrocyte is as phenotypically different from the chondroblast as the precursor to the chondroblast. When we learn more of what induces the dedifferentiation of chondrocytes, we should learn more of what happens when the precursor to the chondroblast is induced to yield a daughter cell which for the first time synthesizes chondroitin sulfate.

References

ABBOTT, J., and H. HOLTZER: Rapid changes in the metabolism of chondrocytes grown *in vitro*. Amer. Zool. **4**, 303 (1964).
— — Critical number of mitoses and differentiation of chondroblasts and myoblasts. Anat. Rec. **151**, 439 (1965).
— — The loss of phenotypic traits by differentiated cells. IV. The reversible behavior of chondrocytes in primary cultures. J. Cell Biol. **28**, 473—487 (1966a).
— — Differences in phenotypic expression of chondrocytes grown in monolayers or in clones. Amer. Zool. **6**, 548 (1966b).
— — Effect of 5-Bromodeoxyuridine on chondrocytes. Proc. Nat. Acad. Sci. **59**, 1144—1151 (1968).
BISCHOFF, R., and H. HOLTZER: The effect of mitotic inhibitors on myogenesis in vitro. J. Cell Biol. **36**, 111—127 (1967).
BEISSON, J., and T. SONNEBORN: Cytoplasmic inheritance of the organization of the cell cortex in *Paramecium aurelia*. Proc. nat. Acad. Sci. (Wash.) **53**, 275—282 (1965).
CAHN, R., and R. LASHER: Simultaneous synthesis of DNA and specialized cellular products by differentiating cartilage cells *in vitro*. Proc. nat. Acad. Sci. (Wash.) **58**, 1131—1138 (1967).
CHACKO, S., and H. HOLTZER: Phenotypic variations in the progeny of a single cartilage cell. Anat. Rec. **160**, 329 (1968).
COON, H. G.: Clonal stability and phenotypic expression of chick cartilage cells *in vitro*. Proc. nat. Acad. Sci. (Wash.) **55**, 66—73 (1966).
COON, H., and R. CAHN: Differentiation *in vitro*: Effects of Sephadex fractions of chick embryo extract. Science **153**, 1116—1119 (1966).
—, and G. MARZULLO: Effect of a factor from embryo extract on cloned chondrocytes. VII. International Congress of Biochemistry. Tokyo, Japan (1967).

D'Abramo, F., and F. Lipmann: The formation of adenosine-3'-phosphate-5'-phosphosulfate in extracts of chick embryo cartilage and its conversion into chondroitin sulfate. Biochim. biophys. Acta (Amst.) **25**, 211—214 (1957).

Grasso, J., and J. Woodard: The relationship between RNA synthesis and hemoglobin synthesis in amphibian erythropoiesis (cytochemical evidence). J. Cell Biol. **31**,279—294(1966).

Henke, K.: Über die verschiedenen Zellteilungsvorgänge in der Entwicklung des beschuppten Flügelepithels der Mehlmotte *Ephestia kühniella Z*. Biol. Zbl. **65**, 120—133 (1946).

—, u. J. Pohley: Differentielle Zellteilungen und Polyploidie bei der Schuppenbildung der Mehlmotte *Ephestia kühniella Z*. Z. Naturforsch. **7**, 65—79 (1952).

Holtzer, H., J. Abbott, J. Lash, and H. Holtzer: The loss of phenotrypic traits by differentiated cells *in vitro*. Proc. Nat. Acad. Sci. **12**, 1533—1542 (1960).

— Aspects of chondrogenesis and myogenesis. In: Synthesis of molecular and cellular structure. Ed. by D. Rudnick, 19th growth symposium, p. 35—87. New York: Ronald Press 1961.

— Comments on induction in cell differentiation. In: Induktion und Morphogenese; 13th Colloq. Gesellsch. Physiol. Chemie, p. 127—143. Berlin-Göttingen-Heidelberg: Springer 1963.

— Regulation of mucopolysaccharide synthesis in the embryo. Biophys. J. **4**,239—250 (1964).

— Induction of chondrogenesis, a concept in quest of mechanisms. In: Epithelial mesenchymal interactions. Ed. by R. Billingham. Baltimore: Williams & Wilkins Co. 1968.

Kawamura, K.: Studies on cytokinesis in neuroblasts of the grasshopper, *Chortophaga viridifasciata*. Formation and behavior of the mitotic apparatus. Exp. Cell Res. **21**, 1—9 (1960).

Lajtha, L.: Cytokinetics and regulation of progenitor cells. J. Cell Physiol. **67**, 133—147 (1966).

Lash, J.: Cell differentiation. Ciba Foundation Symp. p. 114 (1967).

— F. Hommes, and F. Zilliken: The *in vitro* induction of vertebral cartilage with a low-molecular weight tissue component. Biochim. biophys. Acta **56**, 313—319 (1962).

Marks, P., and K. Kovach: Development of mammalian erythroid cells. In: Current topics in developmental biology. Ed. by Moscona, A., and A. Monroy. Vol. 1, pp. 213—252. New York: Academic Press (1966).

Moscona, A., and B. Garber: Keratinizing cells *in vitro*. In: Epithelial-mesenchymal interactions. Ed. by R. Billingham. Baltimore: Williams & Wilkins Co. (in press).

McCulloch, E., J. Till, and L. Siminovitch: Genetic factors affecting the control of hemopoiesis. Canad. Cancer Res. Conf. **6**, 336—356 (1966).

Nameroff, M., and H. Holtzer: The loss of phenotypic traits by differentiated cells. IV. Changes in polysaccharides produced by dividing chondrocytes. Develop. Biol. **16**, 250—281 (1967).

Okazaki, K., and H. Holtzer: An analysis of myogenesis *in vitro* using fluorescein-labeled antimyosin. J. Histochem. Cytochem. **13**, 726—739 (1965).

— — Myogenesis: fusion, myosin synthesis and the mitotic cycle. Proc. nat. Acad. Sci. (Wash.) **56**, 1484—1490 (1966).

Prockop, D., O. Pettingill, and H. Holtzer: Incorporation of sulfate and synthesis of collagen by cultures of embryonic chondrocytes. Biochim. biophys. Acta (Amst.) **83**, 189—196 (1964).

Stebbins, G., and S. Shah: Developmental studies of cell differentiation in the epidermis of monocotyledons. II. Cytological features of stomatal development in the *gramnieae*. Develop. Biol. **2**, 477—500 (1960).

Stockdale, F., J. Abbott, S. Holtzer, and H. Holtzer: The loss of phenotypic traits by differentiated cells. II. Behavior of chondrocytes and their progeny *in vitro*. Develop. Biol. **7**, 293—302 (1963).

—, K. Okazaki, M. Nameroff, and H. Holtzer: 5'-Bromodeoxyuridine: Effects on myogenesis *in vitro*. Science **146**, 533—535 (1964).

Stossberg, M.: Die Zellvorgänge bei der Entwicklung der Flügelschuppen von *Ephestia kühniella Z*. Z. Morph. Ökol. Tiere **34**, 173—206 (1938).

Wessells, N. K.: DNA synthesis, mitosis, and differentiation of pancreatic acinar cells *in vitro*. J. Cell Biol. **20**, 415—433 (1964).

Wilson, D.: Cell in development and heredity. New York: MacMillan 1924.

Zilliken, F.: Notochord induced cartilage formation in chick somites. Intact tissue *versus* extracts. Exp. Biol. Med. **1**, 199—212 (1967).

Phenotypic Expression and Differentiation: *in vitro* Chondrogenesis*

James W. Lash**

*Department of Anatomy, School of Medicine,
University of Pennsylvania, Philadelphia, Pennsylvania, USA*

Any *in vitro* analysis of differentiation should take into account three important factors. One is the artifactual nature of the *in vitro* conditions. The very nature of the culture methodology creates an artifact. This is not meant to imply a deprecation of the study of a created artifact, for in many instances much useful information has been obtained from such studies. The investigator must be aware as to how much transference may be given to *in vivo* phenomena from *in vitro* studies. Another factor to take cognizance of is that negative aspects of differentiation *in vitro* (i. e. "dedifferentiation", or the failure of differentiation) may not be a manifestation of a basic mechanism of differentiation. Observations of this sort may usually be ascribed to the fault of the experimenter and the conditions given the tissue in its foreign environment. A third point to be considered is that the phenotypic expression of differentiation may not be obvious to the observer and may require refined techniques of assay. A cell or tissue that appears "undifferentiated" may actually possess a differentiated metabolic pattern.

Recent reports have exemplified one or more of these points in the works of Przybylski and Blumberg (1966) on muscle, Coon (1966) on chondrocytes, Cahn and Cahn (1966) on retina pigment cells, and Rutter et al. (1967) on pancreas. Using various techniques these reports have shown that the phenotypic appearance of a cell may belie its differentiative capabilities.

This chapter will attempt to show how these factors have played a role in furthering our knowledge of *in vitro* and *in vivo* chondrogenesis. The facts pertaining to vertebral chondrogenesis have been recently reviewed by Lash (1963, 1968) and Strudel (1967). In all vertebrates thus far studied: fish, amphibian, chick, mammal (Watterson 1952; Holtzer 1952; Strudel 1953; Grobstein and Parker 1954) the development of the cartilaginous vertebrae has been shown to be dependent upon the presence of the embryonic spinal cord and notochord. *In vivo* deletion or transplantation experiments (Holtzer 1952; Watterson 1952) have clearly shown this relationship. Excising the embryonic spinal cord creates neural arch deficiencies whereas trans-

* Supported by Grant HD-00380 from the National Institute of Child Health and Human Development, National Institutes of Health, and National Science Foundation Grant GB-3674 and GB-6748.

** Career Development Awardee of the National Institute of Child Health and Human Development.

plantation of a piece of spinal cord into the lateral somitic region induces supernumerary neural arch cartilage. The same type of inductive interactions may be
surmised from human congenital malformations such as the myeloschises resulting
in spina bifida. In these instances there occurs a disturbance in the normal interactions
between the spinal cord and adjacent somitic mesenchyme, resulting in abnormal
neural arches (LASH 1964).

Most of the work on these interactions have been performed using *in vitro* mani-
pulations of embryonic chick tissues. Embryonic chick somites of $2^1/_2$ to 3 days incubation will form little or no cartilage if explanted to tissue culture. If the somites
are cultured in association with the embryonic spinal cord or notochord, all of such
cultures form significant amounts of cartilage matrix. The cartilage appears on the
third or fourth day of culture, approximately the same time it would appear if the
somites were *in vivo*. Recent work has shown that hypertrophied cartilage from
embryonic limbs or ribs (COOPER 1965) or extracts from the spinal cord or notochord
(STRUDEL 1962; LASH, HOMMES and ZILLIKEN 1962; ZILLIKEN 1967) can also promote chondrogenesis *in vitro*.

The spinal cord and notochord can exert their chondrogenic stimulus through a
millipore filter, indicating that a transmissable substance is involved (LASH, HOLTZER,
and HOLTZER 1957; COOPER 1965; FLOWER and GROBSTEIN 1967). Additional evidence for a transmissable substance was originally obtained from conditioned nutrient
agar experiments. In these experiments 0.2 mm segments of three day embryonic
chick spinal cord were cultured on nutrient agar for 24 hours, then removed and
replaced by clusters of somites. These somite clusters formed significantly more
cartilage than did somites placed elsewhere in the same agar dish. Since the prospect
of extracting a cartilage promoting agent from nutrient agar seemed too formidable,
these experiments led to the attempts to extract spinal cords and notochords (LASH,
HOMMES and ZILLIKEN 1962; ZILLIKEN 1967).

The mechanism by which explanted somites become stimulated to form cartilage
is still unknown, though we now have more clues concerning the somites' response
to promoting factors or tissues. It is possible that the promoting agents prevent the
escape of necessary metabolites from the somites. This seemed likely from the results
obtained correlating spontaneous chondrogenesis with the mass of the somite cluster
(HOLTZER 1964). In these experiments the larger the mass of somites, the greater the
incidence of chondrogenesis. This was one of a number of experiments indicating
that somites under certain culture conditions could form cartilage without exogenous
stimulation, i.e. spontaneous cartilage formation (cf. LASH, GLICK, and MADDEN
1964). As culture methods improved, the incidence of spontaneous cartilage increased.
The increasing frequency of spontaneous cartilage formation prompted a study to
correlate chondrogenic bias with the histology and age of the somites (LASH, GLICK,
and MADDEN 1964; LASH 1967). It became obvious that the histoarchitecture of the
isolated somite was strongly correlated with subsequent cartilage formation. The
more anterior somites, with increasing differentiation of myotomic, dermatomic, and
sclerotomic regions, formed significantly less cartilage than did the less differentiated
posterior somites from the same animal. At this stage the posterior somites consist
primarily of an epithelial ball containing a core of mesenchymal cells (Fig. 1). Why
these latter somites display such greater chondrogenic bias in culture is unknown.
It may be correlated with the fact that the preparative procedures of isolation damage

the sclerotomal cells of the anterior somites to a point where they can not recover in the absence of exogenous stimulation. Indeed, it has been shown that after the trypsinization procedure during preparative dissection of the embryo, the total protein content of the embryo is markedly reduced, and the specific enzyme activity associated with the sulfokinase system is seriously impaired (LASH, GLICK, and MADDEN 1964). Microscopic observations show that trypsinized anterior somites are more seriously disrupted than the posterior ones.

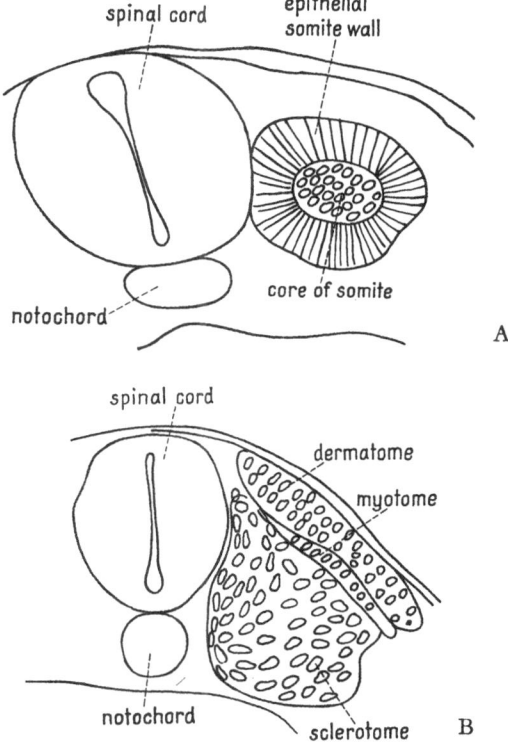

Fig. 1. Diagrams of sections through the anterior (1A) and posterior (1B) regions of a 2½ day chick embryo. The advanced differentiation in 1B is seen as discernible dermatome, myotome, and sclerotome. The posterior somites at this stage consist of an epithelial ball containing core cells. (cf. LASH, 1967)

With this clear difference between the behavior of the anterior and posterior somites established, experiments were performed to test the notion that an increase in mass alone was sufficient to provoke chondrogenesis. For these experiments the somites were divided according to their position in the body (anterior − adjacent to the anterior limb primordia, middle − the region between the two limb primordia, and posterior − adjacent to the posterior limb primordia). It was seen (Table 1) that regardless of the mass, i.e. the number or size of the somites in a cluster, the ability of the somites to form cartilage was correlated with their position in the embryo. The anterior somites formed little cartilage regardless of the size of the mass, and the

2*

posterior somites formed significant amounts, regardless of the mass. If the somites were mixed up and randomized so that anterior, middle and posterior were in the same cluster, then the mass effect was observed. The increased amount of cartilage however, could definitely be correlated with the increased probability of including posterior somites in the larger clusters.

Table 1

Somite Clusters	No. Somites	Incidence of Cartilage at		
		4 days	6 days	8 days
Anterior	6	0/15	2/15	4/15
	12	0/15	0/15	1/15
	18	0/15	1/15	3/15
Middle	6	0/12	2/12	7/12
	12	0/10	2/10	7/10
	18	0/10	2/10	8/10
Posterior	6	8/10	8/10	10/10
	12	4/10	8/10	10/10
	18	4/10	6/10	10/10

The incidence of cartilage in cultures of somite clusters from $2^1/_2$ day chick embryos. The appearance of cartilage is correlated with the region of the embryo, and the incidence is the same within each group (anterior, middle, posterior), regardless of the size of the somite mass. (From LASH 1967, 1968)

Thus, the somites have a strong chondrogenic bias at the time they are explanted. The effect, therefore, of the notochord or other cartilage promoting agents, is not to impart the ability to form cartilage, but rather to enhance the existing chondrogenic proclivity. Indeed, a recent report by ELLISON and AMBROSE (1967) indicates that an enriched medium is sufficient to permit spontaneous cartilage formation in explanted somites. An interesting difference between spontaneously formed cartilage and that promoted by external stimuli (e.g. "induced") should be noted. In the former, the continued presence of the enriched medium is necessary, whereas just a few hours association with the spinal cord is sufficient for induced cartilage formation. The spinal cord, and presumably the notochord also, may be removed from its association with the somites after a few hours, and the somites will still form cartilage 3—4 days later (LASH, HOLTZER, and HOLTZER 1957). It is not possible to say whether these two instances represent different biochemical aspects of chondrogenesis. If the biochemical initiation of chondrogenesis may seem to differ, the response is the same in that the somites accumulate cartilage matrix.

More compelling evidence for the chondrogenic bias of embryonic somites comes from a variety of sources. JOHNSTON and COMAR (1957) have shown that chick embryos selectively incorporated radioactive sulfate into regions associated with future chondrogenesis (somitic mesenchyme around the spinal cord and notochord). Later work by FRANCO-BROWDER, DE RYDT, and DORFMAN (1963) reported the isolation of chondroitin sulfate from embryonic chicks well before matrix appeared. LASH, GLICK, and MADDEN (1964) also obtained sulfate incorporation coinciding with a metachromatic staining material in the region bordering the spinal cord and notochord, in the somites, and in the subepidermal region.

More exacting information on the chondrogenic potential of the somites was obtained with the development of a thin-layer chromatographic system to separate various glucosamine metabolites and the chondroitin sulfates (MARZULLO and LASH 1967; LASH 1968).

Glucosamine
  ↓ ATP
Glucosamine-6-P
   ⇵ acetyl-CoA
N-acetylglucosamine-6-P
   ⇵
N-acetylglucosamine-1-P   ↗ Hyaluronic acid
   ⇵ UTP
UDP-N-acetylglucosamine (UDP-NAG) ⟶ Keratosulfate
   ⇵ (UDP-NAG-4-epimerase)
UDP-N-acetylgalactosamine (UDP-NAGal) ↘ Heparitin sulfate
   ↓ UDP-glucuronate
Chondroitin
   ↓ PAPS
Chondroitin sulfates

Fig. 2. Glucosamine metabolism leading to the formation of chondroitin sulfate. ^{14}C-glucosamine was added to tissues and the various intermediates isolated by thin-layer chromatography (From MARZULLO and LASH, 1968)

The chondroitin sulfates are the only normal connective tissue components to utilize UDP-NAGal (cf. Fig. 2), hence the abilities of tissues to metabolize glucosamine and to form UDP-NAGal would be indicative of a chondrogenic genotype. The assay for this may be done in two ways. One is to demonstrate the formation of UDP-NAGal in a metabolizing system, the other is to assay for the activity of the enzyme UDP-NAG-4-epimerase (Fig. 2). With the histological and biological evidence already at hand, it was no surprise that the isolated somites readily metabolized glucosamine to form UDP-NAGal, and in general possessed a metabolic pattern similar to that of differentiated cartilage (MARZULLO and LASH 1967). The cartilage-type metabolic activity was also found in all other embryonic tissues (epidermis, spinal cord, notochord, endoderm, mesonephros, extra-embryonic membranes and lateral mesenchyme). The metabolism of glucosamine in these other tissues was indistinguishable from that in somites (MARZULLO and LASH 1967). All embryonic tissues, however, including the somites, differed in one significant aspect from the differentiated cartilage. Embryonic tissues accumulate more of the hexosephosphates in an incubation with a labeled precursor (^{14}C-glucosamine) whereas cartilage accumulates more of the terminal products, the UDP-NA-hexosamines (see Fig. 2 for pathways; MARZULLO and LASH 1967). Thus with respect to glucosamine metabolism, embryonic somites possess the same enzymes as mature cartilage, but are less efficient in metabolizing N-acetylglucosamino-phosphates to the UDP-NA-hexosamines. This particular step in the metabolic pathway is under further investigation to determine what role it may play in the stabilization or enhancement of the chondrogenic bias.

Comparable studies are in progress with the sulfate-activating enzymes. These enzymes (ATP-sulfurylase and APS-kinase) are responsible for the formation of "active sulfate" (PAPS, 3'-phosphoadenosine-5'-phosphosulfate) (LIPMANN 1958).

PAPS, through a transfer enzyme, is the source of the sulfate for the glycosamino-glycans of the cartilage matrix (chondroitin sulfates A and C, and keratosulfate). Using assay methods less sensitive than those used for glucosamine metabolism it was evident that the early embryonic tissues possessed the sulfokinase enzymes as well as the transfer enzymes (GLICK, LASH, and MADDEN 1964). The activity of these enzymes was not detectable after trypsinization.

More crucial evidence for the synthesis of tissue-specific macromolecules by early morphologically unspecialized tissues was obtained from micro assays for the production of chondromucoproteins (MARZULLO and LASH 1967). Thin-layer chromatographic techniques permitted the isolation and characterization of chondroitin sulfate and keratosulfate from the embryonic tissues. An exhaustive survey is presently in progress, but results indicate that the embryonic somites can synthesize chondromucoprotein. There is little doubt that the "pre-induced" somites have all of the metabolic machinery necessary for producing chondromucoprotein. They do not, however, accumulate the product *in vitro* unless various other conditions are met ("induction", enriched medium). What was implicitly assumed in the past to be an instance of tissue induction with *de novo* synthesis of matrix, is now shown not to be the induction of a new synthetic pathway, but the stabilization and enhancement of a pre-existing one. The term induction may still apply to this situation, but being an operational definition, it now assumes a different meaning for this interaction system. The stabilization of the chondrocytic metabolic pattern is produced *in vitro* and the result is the accumulation of matrix and easily detectable and visible chondrogenesis. Whether this type of induction operates in other interacting systems is not known, but there are reports indicating that some differentiating systems may operate similarly.

Although the initiation of limb chondrogenesis is not recognized as an interacting system, it has been shown that the limb mesenchyme not only has the necessary enzymes for chondroitin sulfate synthesis well before the time of chondrogenesis (MEDOFF 1967), but also synthesizes it in detectable quantities (SEARLS 1965). As in the somites, they do not accumulate the product during the early developmental stages, but make it at barely perceptible levels. When frank chondrogenesis occurs, the specific activity of the enzyme increases (MEDOFF 1967) and matrix accumulates. It would appear evident that this low-level and non-accumulative synthetic activity in the limb-buds is similar to the activity found in all other embryonic tissues of comparable age. Because of the ubiquity of connective tissue elements throughout the organisms, and the similarity of connective tissue synthetic patterns, caution must be taken before a generalized scheme of differentiation is approached. It is tempting to ascribe this low-level, non-accumulative synthetic activity to the fact that in any unspecialized tissue during early development there occurs a greater range of potentialities (i.e. informational molecules) than after tissue specialization. From the work on ascorbic acid synthesis (FABRO and RINALDINI 1965), hemoglobin (WILT 1962) and pancreas (RUTTER 1967), specialized molecules may be synthesized extremely early in development.

This brings us back to an older problem in embryology, one that is still unresolved, and that is, what is the origin of determination and competence? There are varying degrees of developmental pliancy in different organisms. In the Ascidians for example, tissue areas appear destined in the cytoplasm of the fertilized egg before cleavage. In the most regulative eggs (e. g. echinoderms) there is little determinance as late as the eight-cell stage, and some tissue areas remain undetermined relatively late in embryonic development. All tissues eventually become determined with a restriction of competence, but do so at different rates in different organisms.

It has taken many years to acquire the meager knowledge we have concerning the biochemical events associated with interacting tissues. These events are too subtle to have been appreciated with older techniques. With a constant improvement in techniques and biochemical knowledge, the problem of embryonic induction, determination, and competence may eventually come to be understood.

References

CAHN, R. D., and M. B. CAHN: Heritability of cellular differentiation: Clonal growth and expression of differentiation in retinal pigment cells *in vitro*. Proc. nat. Acad. Sci. (Wash.) **55**, 106—114 (1966).

COON, H. G.: Clonal stability and phenotypic expression of chick cartilage cells *in vitro*. Proc. nat. Acad. Sci. (Wash.) **55**, 66—73 (1966).

COOPER, G. W.: Induction of somite chondrogenesis by cartilage and notochord: A correlation between inductive activity and specific stages of cytodifferentiation. Develop. Biol. **12**, 185—212 (1965).

ELLISON, M., and E. J. AMBROSE: Manuscript (1967).

FABRO, S. P., and L. M. RINALDINI: Loss of ascorbic acid synthesis in embryonic development. Develop. Biol. **11**, 468—488 (1965).

FLOWER, M., and C. GROBSTEIN: Interconvertibility of induced morphogenetic responses of mouse embryonic somites to notochord and ventral spinal cord. Develop. Biol. **15**, 193—205 (1967).

FRANCO-BROWDER, S., J. DE RYDT, and A. DORFMAN: The identification of a sulfated mucopolysaccharide in chick embryos. Stages 11—23. Proc. nat. Acad. Sci. (Wash.) **49**, 643—647 (1963).

GLICK, M. C., J. W. LASH, and J. W. MADDEN: Enzymic activities associated with the induction of chondrogenesis *in vitro*. Biochim. biophys. Acta (Amst.) **83**, 84—92 (1964).

GROBSTEIN, C., and G. PARKER: *In vitro* induction of cartilage in mouse somite mesoderm by embryonic spinal cord. Proc. Soc. exp. Biol. (N. Y.) **85**, 477—481 (1954).

HOLTZER, H.: An experimental analysis of the development of the spinal column. I. Response of pre-cartilage cells to size variations of the spinal cord. J. exp. Zool. **121**, 121—148 (1952).

— Control of chondrogenesis in the embryo. Biophys. J. Suppl. **4**, 239—250 (1964).

JOHNSTON, P. M., and C. L. COMAR: Autoradiographic studies of the utilization of S^{35}-sulfate by the chick embryo. J. biophys. biochem. Cytol. **3**, 231—238 (1957).

LASH, J. W.: Tissue interaction and specific metabolic responses: Chondrogenic induction and differentiation. In: Cytodifferentiation and macromolecular synthesis. Ed. M. LOCKE, p. 235—260. New York: Academic Press 1963.

— Normal embryology and teratogenesis. Amer. J. Obstet. Gynec. **90**, 1193—1207 (1964).

— Differential behavior of anterior and posterior embryonic chick somites *in vitro*. J. exp. Zool. **165**, 47—56 (1967).

— Somitic mesenchyme and its response to cartilage induction. In: Epithelial-mesenchymal interactions. Ed. R. FLEISCHMAJER and R. BILLINGHAM, in press. Baltimore: Williams & Wilkins 1968.

—, S. HOLTZER, and H. HOLTZER: An experimental analysis of the development of the spinal column. VI. Aspects of cartilage induction. Exp. Cell Res. **13**, 292—303 (1957).

—, F. A. HOMMES, and F. ZILLIKEN: Induction of cell differentiation. The *in vitro* induction of vertebral cartilage with a low-molecular weight tissue component. Biochim. biophys. Acta (Amst.) **56**, 313—319 (1962).

—, M. C. GLICK, and J. W. MADDEN: Cartilage induction *in vitro* and sulfate-activating enzymes. Nat. Canc. Inst. Monogr. **13**, 39—49 (1964).

Lipmann, F.: Biological sulfate activation and transfer. Science **128**, 575—580 (1958).

Marzullo, G., and J. W. Lash: Acquisition of the chondrocytic phenotype. In: Exp. Biol. Med., Vol. 1, pp. 213—218. Basel-New York: S. Karger 1967.

Medoff, J.: Enzymatic events during cartilage differentiation in chick embryonic limb bud. Develop. Biol. **16**, 118—143 (1967).

Przybylski, R., and J. M. Blumberg: Ultrastructural aspects of myogenesis in the chick. Lab. Invest. **15**, 836—863 (1966).

Rutter, W. J., W. D. Ball, W. S. Brandshaw, W. R. Clark, and T. G. Sanders: Levels of regulation in cytodifferentiation. In: Exp. Biol. Med. Vol. 1, pp. 110—124. Basel-New York: S. Karger 1967.

Searls, R. L.: Isolation of mucopolysaccharide from the pre-cartilaginous embryonic chick limb bud. Proc. exp. Biol. (N. Y.) **118**, 1172—1176 (1965).

Strudel, G.: Influence Morphogene du Tube Nerveux et de la chorde sur la Differenciation de la Colonne Vertebrale. C. R. Soc. Biol. (Paris) **147**, 132—133 (1953).

— Induction de Cartilage *in vitro* par L'extrait du Tube Nerveux et de la Chorde de L'embryon de Poulet. Develop. Biol. **4**, 67—86 (1962).

— Some aspects of organogenesis of the chick spinal column. In: Exp. Biol. Med., Vol. 1, pp. 183—198, Basel-New York: S. Karger 1967.

Watterson, R. L.: Neural tube extirpation in *Fundulus heteroclitus* and resultant neural arch defects. Biol. Bull. **103**, 310 (1952).

Wilt, F.: The ontogeny of chick embryo hemoglobin. Proc. nat. Acad. Sci. (Wash.) **48**, 1582—1590 (1962).

Zilliken, F.: Notochord induced cartilage formation in chick somites. Intact tissues versus extracts. In: Exp. Biol. Med., Vol. 1, pp. 194—212. Basel-New York: S. Karger 1967.

The Nature and Probable Cause of Modulations in Pigment Cell Cultures*

J. R. Whittaker**

Department of Zoology, University of California, Los Angeles, California

I. The Problem

Melanotic pigment cells have been known for many years to lose pigmentation traits in cell and tissue culture. Pigment cells from the iris and retinal pigment layer (tapetum) of chick embryos and also neoplastic melanoma cells derived from mammalian skin melanocytes have all been observed to do this, and frequently they will reacquire pigmentation under altered culture conditions (examples reviewed by Whittaker 1963, 1967). The loss of melanotic phenotype in cultures of pigment cells and the possible reappearance of pigmentation after many cell generations is one of the refreshingly simpler examples in cell culture of what can be considered a modulation in the terminology of Weiss (1949). Modulation is a reversible fluctuation within the already established range of determination; it is also the covert maintenance of the potentiality to express an original function. Although many kinds of cells *in vitro* undergo changes of phenotypic expression, relatively few cell modulations have been examined closely. Phenotypic changes in cell culture have an intrinsic interest and fascination as manifestations of unusual cell behavior, but the modulation phenomenon also appears to be related to the important problems of cell maintenance and stability.

The fundamental control of cell activity is genetic. Genes specify the directions in which cells will differentiate, and they presumably regulate the stability of the differentiated state once it is achieved. The question is, how close a control do the genes exert over cell phenotype? Many physiological changes occur which are not genetically controlled in the immediate sense. One would like to know if modulations involve a transitory change in activity of genes concerned with specific phenotypes. For practical purposes the active synthesis of a specific protein (gene expression) can be usually examined and equated with gene activity. The answer to the genetic question, as suggested by the work reviewed here, is that melanotic pigment cells are capable of losing specific gene expression for reasons not directly related to the regulation of gene activity.

* Supported by research grant HD-00059 from the National Institute of Child Health and Human Development, U.S. Public Health Service, and by Cancer Research Funds of the University of California.

** Present affiliation: The Wistar Institute of Anatomy and Biology. Philadelphia, Pennsylvania.

II. Pigment Cell Changes in Culture

A. Loss of Phenotypic Traits

Embryonic chick retinal pigment cells when dissociated from the tapetal tissue and grown in monolayer culture lose both morphological and biochemical characteristics of their melanotic state of differentiation. In an earlier study (WHITTAKER 1963), tissues from $5^1/_2$-day old embryos (Hamburger-Hamilton stage 28) were used, but in subsequent experimental work (WHITTAKER 1967, 1968) tissues of $6^1/_2-7$ day embryos (stages 30 and 31) were studied; within the age range of $5^1/_2-7$ days, age of tissue was not a factor in the behavior of the cells in culture.

The cells were cultured in an embryo extract-containing medium with equal parts of 12-day embryo extract, horse serum, and Tyrode's solution. In organ culture experiments, where aggregates of cells from previous monolayer cultures were maintained at the air-medium interface on Millipore filter rafts, the same embryo extract-containing medium was used. Details of the cell culturing procedures will be found in WHITTAKER (1963, 1967).

Diminished melanin content can be observed visually in cells recovered after different periods in monolayer culture. When this change is measured quantitatively by optical density measurements of extracted and dissolved cells, the melanin content on a per unit protein basis in equal cultures is essentially the reciprocal of the increase in total protein per culture in the same culture series (WHITTAKER 1967). The quantitative change observed is a dilution of pigment due to cell growth (defined in this and subsequent discussion as increase in total protein); there is no actual loss of melanin on a per culture basis. Thus disappearance of melanin is not an active loss or extrusion of melanin granules. Measurements of melanin change in culture indicate that there is little increase in total melanin per culture with time. When the change in the melanin content of cells in culture for 12 hours (0−12 hours) was compared with the change in the melanin content of equivalent tissues during 12 hours *in ovo*, the comparison pointed to a drastic reduction in melanin accumulation during culture *in vitro*.

Chick retinal pigment cells (and chick embryo cells generally) are not capable of indefinite propagation *in vitro*, but with moderate care and frequent subculturing they were maintained as growing cells in the unpigmented state for up to 3 months. During this time the cells retained their epithelioid properties and readily formed continuous epithelial monolayer sheets following repeated subculturing.

The loss of melanotic characteristics is a modulation and depigmented cells are able to become pigmented once again if crowded monolayer cultures are left for a week or more without subculturing, or if cells taken from growing monolayer cultures are crowded together in organ cultures (WHITTAKER 1963). Such redifferentiation experiments are reproducible, but difficult to work with quantitatively. Consequently most of what will be reviewed in this report about pigment cell modulation is based on analysis of the *loss* of phenotypic characters.

B. Changes in Melanin Biosynthesis

The rapid decrease in melanin synthesis suggested by measurements of changes in the melanin/protein ratio has been investigated further by examining the rate of incorporation of radioactive tyrosine-2-^{14}C into the melanin of pigment cells during culture.

This was measured as the tyrosinase-dependent incorporation of the amino acid using a non-toxic tyrosinase inhibitor. The technique and its validity has been discussed at length previously (WHITTAKER 1966, 1967). Pigment cells or intact pigment tissue pieces continue to synthesize melanin at a very high rate during the first few hours (0—4 hours) of culture. Thereafter, the rate of melanin biosynthesis drops rapidly, until during 24—36 hours of culture it has reached a relatively low level (Fig. 1). Measurements of total protein changes in similar culture series show that in 36 hours of growth in culture the protein content of the cultures has approximately doubled. However, the rate of melanin biosynthesis by 36 hours of culture is lower than would be expected on the basis of this simple dilution. Other changes must be contributing to the loss of activity.

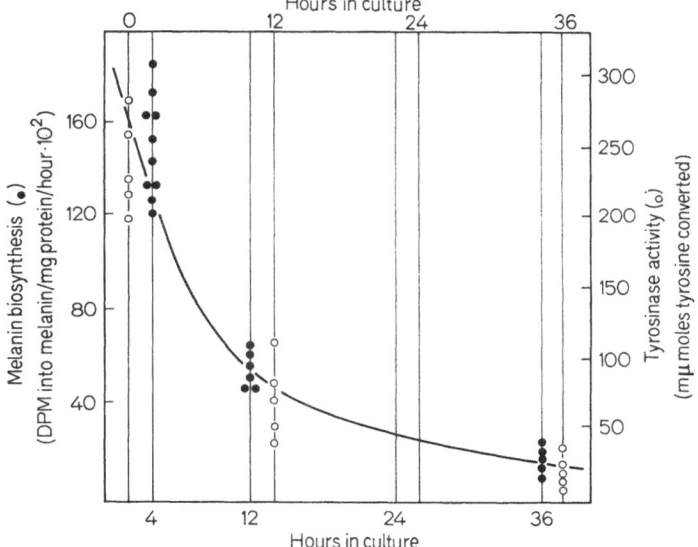

Fig. 1. Rate of melanin biosynthesis, and tyrosinase activity in chick retinal pigment cell cultures (from WHITTAKER 1967). The periods of melanin biosynthesis assayed were 0 to 4 hours, 0—12 hours, and 24—26 hours. Both sets of measurements are expressed on a per unit protein basis

This test of the rate of melanin biosynthesis is apparently also faithfully measuring the *in vivo* tyrosinase activity. If tyrosinase is assayed by a direct measurement of enzyme activity in homogenates made at different times of culture (WHITTAKER 1967), one obtains an activity pattern virtually identical to the pattern of changes in melanin biosynthesis (the superimposed results of both kinds of analysis are shown in Fig. 1). In addition to dilution, changes in the synthesis and activity of tyrosinase are probably the basis of the modulation phenomenon shown by pigment cells.

C. Tyrosinase Synthesis and Stability

An attempt was made to pinpoint the time at which tyrosinase synthesis occurs in the pigment cell cultures. Inhibition of 70—80% of new protein synthesis with the antibiotic cycloheximide lowered the measurable level of tyrosinase activity significantly (WHITTAKER 1967). When homogenates were tested following cycloheximide treatment for the 24—36 hour culture period, there was no effect of the drug on enzyme level. Further experiments to determine the time of synthesis, using the effect of cycloheximide on 3—15 hour and 12—24 hour cultures, indicated that there was enzyme synthesis in the first of these two periods but not during the second (WHITTAKER 1968). Thus, tyrosinase synthesis does not occur after 12 hours of cell culture. Since there is synthesis during the first 12 hours of culture, one would expect the levels of tyrosinase activity and melanin biosynthesis to fall more slowly than they do (Fig. 1).

Utilizing the observation that tyrosinase was not synthesized after 24 hours, 24-hour old cultures containing equivalent numbers of cells were assayed for tyrosinase activity at subsequent 24-hour intervals to see if the total enzyme per culture was disappearing at some constant rate (WHITTAKER 1967). In fact, the results showed that the enzyme activity decreased in a perfectly exponential fashion, as one would predict for decay (Fig. 2 A). The rate of decay thus measured (40% of residual activity per 12 hour period in culture), if constant for earlier times of culture as well, is sufficient in conjunction with the growth dilution to explain the rapid drop in enzyme activity and melanin biosynthesis observed (Fig. 1). Decay of the enzyme is probably due to a reaction inactivation (wearing out). If cells are reared for the initial 24 hours of culture at 37° C, and then transferred to a continuous temperature of 22° C, the tyrosinase activity decays only very slightly (Fig. 2 B). Apparently the enzyme is more stable in less metabolically active cells, which supports the idea that decay is related to enzyme function.

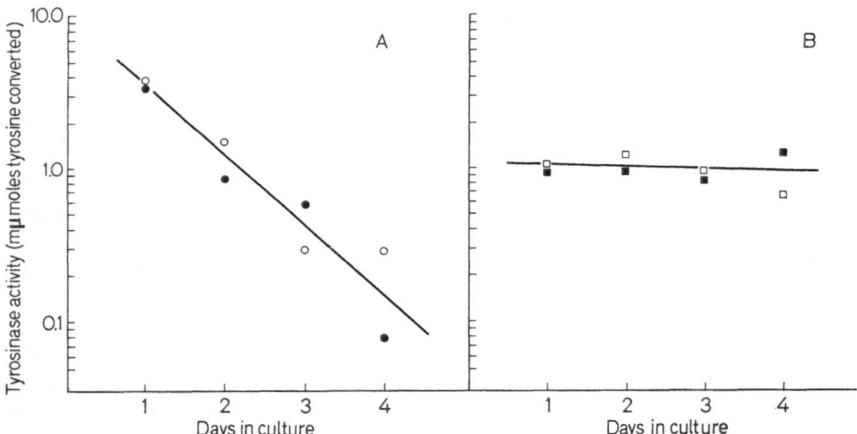

Fig. 2. Tyrosinase activity in different series of chick retinal pigment cell cultures grown at 37° C (A), and at 22 °C (B) following an initial 24 hours of culture at 37° C. The enzyme activity of homogenates is expressed on a per culture basis. Data in (A) are from WHITTAKER (1967); results in (B) were previously unpublished

An important point to pursue is whether tyrosinase synthesis stops completely after 12 hours, or continues indefinitely at an extremely low level. The melanin biosynthesis and tyrosinase activity assays used (WHITTAKER 1963, 1967) have been applied to older cultures where they *do* indicate low activity levels, but unfortunately these assay techniques are not particularly sensitive and reliable at really low levels. Previous observation of a few small melanin granules in very old cultured cells suggests that a low level of tyrosinase synthesis must occur (WHITTAKER 1963). Recently, a more sensitive biosynthesis assay with dihydroxyphenylalanine-2-^{14}C as a substrate *in vivo* has been perfected and used to demonstrate that a small residue of melanin biosynthesis (and presumably tyrosinase synthesis) still occurs in pigment cells after 21 days in monolayer culture (WHITTAKER 1968). The implication is that loss of tyrosinase synthesis is a quantitative rather than qualitative change. This may explain why resumption of melanin synthesis occurs so readily under altered culture conditions; it has never completely stopped.

Thus three changes seem to describe the modulation phenomenon in pigment cells. Cell growth dilutes the components of the melanogenic system, tyrosinase activity decays, and tyrosinase synthesis diminishes to an ineffectively low level which cannot compete with either the decay or the dilution. All three changes occur quite rapidly. If regulation of gene activity is to be considered as an explanation of modulation, then the change in tyrosinase synthesis is the most important of the three.

There is an additional indication that regulation of tyrosinase synthesis may be at the basis of the problem. Other proteins which are apparently involved in melanogenesis continue to be synthesized after tyrosinase synthesis has effectively stopped. This is shown by the effect of both cycloheximide and actinomycin D (an inhibitor of new RNA synthesis) on melanin biosynthesis. Both inhibit melanin biosynthesis during 24—36 hours, a time when neither has any effect on the level of tyrosinase activity. This inhibitor-sensitive synthesis might represent the granule components postulated by MOYER (1966) but it may also be proteins which have other cell functions as well. Structural protein components of the melanin granule (melanosome) undoubtedly exist, and melanin is thought to be synthesized as a melanoprotein. However, there is overwhelming evidence from work with another animal cell system that melanin does not occur *in situ* as a melanoprotein (WHITTAKER 1966).

III. Possible Causes of the Diminished Tyrosinase Synthesis

A. Cell Selection

The fact that retinal pigment tissue (tapetum) is a histologically pure tissue type consisting solely of melanotic pigment cells negates the obvious argument about selection of a nonpigmented cell type under culture conditions as the cause of phenotypic change (SATO, ZAROFF, and MILLS 1960). There might nevertheless be selection of certain tapetal cells which are deficient in pigment synthesizing ability. However, attachment of disaggregated cells to the culture substratum can be as high as 60% in some experiments (WHITTAKER 1967), and there is also clear evidence that almost all of these attached cells are dividing in culture. Radioautographic examination of thymidine-^{3}H incorporation into cell nuclei indicates that at 36 hours, prior to actual first cell division, virtually all of these attached cells are making DNA (WHITTAKER 1968). In addition, examination of the culture medium at 12 or 36 hours, after the initially unattached cells have been removed (at 4 to 6 hours), shows that very few cells are lost into the medium.

Selection of abnormal karyotypes, possibly lacking essential genes, has occasionally been invoked as an explanation of phenotypic changes in cell culture (GRANT

1965). However, two features of chick retinal pigment cell behavior contradict this hypothesis. The major phenotypic change takes place prior to first cell division. Abnormal karyotypes would not be expected to occur before first mitosis and several generations of abnormal mitoses would be necessary to produce a variety of deficient karyotypes. Secondly, the reversible nature of pigment cell dedifferentiation, even after many generations in culture is inconsistent with an hypothesis requiring loss of genetic material concerned with pigment synthesis.

B. An Environmental Substance

If diminished tyrosinase synthesis is caused directly by some physical or chemical property of the culture environment then the effect must be a rather specific one. The rate of overall protein synthesis is rising rapidly in culture, as measured by leucine-^{14}C and valine-^{14}C incorporation (WHITTAKER 1968), so that synthesis of many specific proteins is not affected. Also, an enzyme essential to general cell function, cytochrome oxidase, has been shown not to change in activity during pigment cell culture (WHITTAKER 1963).

Although tyrosinase requires the amino acid tyrosine as a substrate for its activity, nothing as simple as the availability of tyrosine explains the enzyme changes. The culture medium already contains a reasonable amount of tyrosine, and the addition of extra quantities of tyrosine does not delay or prevent the loss of melanotic phenotype (WHITTAKER 1963). There are inhibitors and activators of tyrosinase which occur in many animal cell systems and possibly the appearance or disappearance of these have a regulatory effect on enzyme activity (for example, see CHIAN and WILGRAM 1967). However, naturally occurring inhibitors and activators are not known to affect synthesis of the enzyme. Since loss of phenotypic traits is of such widespread occurrence in different kinds of cultured cells some more general mechanism is likely to be the cause of enzyme changes in culture.

All hypotheses that assume the presence (or absence) in the culture medium of some substance which suppresses differentiation must account for the observation that dedifferentiated pigment cells taken from monolayer cultures and grown in organ culture as a reaggegated tissue mass produce melanin pigment when maintained in medium of composition identical to that used for the monolayer cultures (WHITTAKER 1963). Fig. 3 shows a section through such a tissue mass of 43-day old dedifferentiated cells, many of which have become repigmented after only 5 days in organ culture. The melanin reappears in some cells as masses of very tiny granules and in other cells as dense diffuse pigment which seems to be dissolved throughout the cell (Fig. 4), but not as the characteristic rodshaped granules seen originally in the cells taken from the embryo.

Perhaps the cells themselves elaborate a substance which is either present or absent depending upon the response of the cells to particular growth or culture conditions. While less complicated and more conservative explanations seem reasonable at the present time, no available information rigorously excludes the possibility of variations in the endogenous production of a nutritional or inhibitory substance.

CAHN and CAHN (1966) and COON and CAHN (1966) have suggested that chick embryo extract contains an inhibitor of differentiation. Their hypothesis is based on the different effects of high and low molecular weight fractions of chick embryo extract on maintenance of differentiation in cells growing as clonal cultures. Chick retinal pigment cells grown clonally by CAHN and CAHN (1966) in medium containing the low molecular weight fraction of chick embryo extract maintain a visible level of pigmentation, whereas similar cultures in medium containing the high molecular weight fraction (larger than molecular weight 5000) lose their visible pigmentation. There is a notable difference in the reported generation times of cultures grown in the two media, with the high molecular weight embryo extract causing more rapid growth. However, cell number change is not a reliable index of growth in a situation where significant changes can occur faster than cells divide. It would be interesting to know what rate of melanin synthesis occurs in the low molecular weight medium compared to the rate of synthesis in fresh intact tissues. Judging from the description of the sequence of pigmentation changes in the clones (CAHN and CAHN 1966) there is probably a considerable quantitative reduction in the activity.

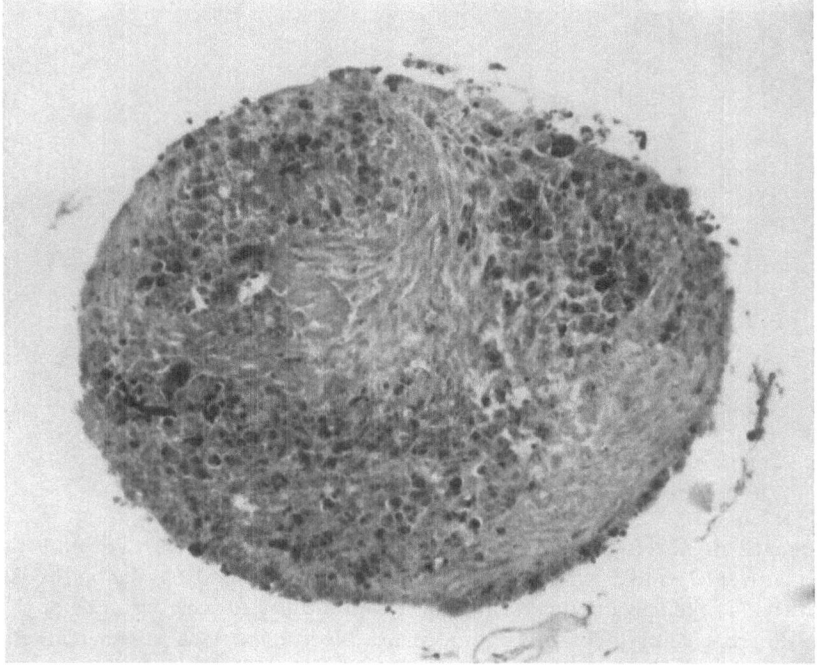

Fig. 3. $6^1/_2$-day chick retinal pigment cells grown for 43 days in monolayer culture and transferred to organ culture for 5 days as an aggregated mass. Fixed in Bouin's, sectioned at 7 μ thickness, and stained in hematoxylin-eosin. A whole aggregate showing 3 large sectors of pigmentation. Magnification: × 150.

The CAHN and COON theory of a differentiation inhibitor in embryo extract loses plausibility with their own demonstration that cultures of higher cell density do not remain differentiated even in low molecular weight medium. In addition, my own observations, illustrated by Fig. 3, show that crowded or densely packed cells in unfractionated embryo extract-containing medium will readily synthesize pigment. This is a puzzling series of observations to be explained by an inhibitor hypothesis, without complicated additional assumptions. There are also sufficient examples of phenotypic loss in media *not* containing embryo extract to raise serious doubts about its validity as a general theory of phenotypic suppression (eg. SATO, ZAROFF, and MILLS 1960; HILFER 1962; PRIEST and PRIEST 1964; TWAROG and LARSON 1964; DANIEL and DE OME 1965). The medium is apparently not the message.

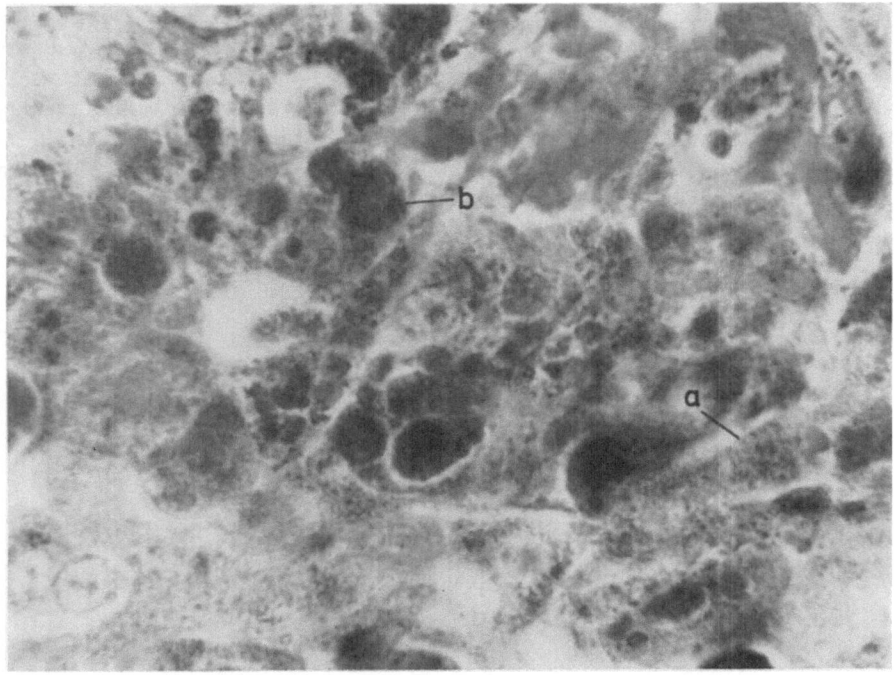

Fig. 4. A pigmented area of the aggregate in Fig. 3 with cells containing tiny granules (a), and cells with dense diffuse pigmentation (b). Magnification: × 1200

Considering the frequently observed antagonism between growth and differentiation both *in vivo* and *in vitro*, a more attractive theory would be based on the assumptions that: (1) the specific qualitative and quantitative growth responses of the cells under different stable growth conditions are more important in the regulation of phenotype than is the nature of the growth initiating stimulus, and (2) these specific growth responses of the cells are in some way related to diminished enzyme synthesis.

C. Growth and Macromolecular Synthesis

As noted above, most synthesis of tyrosinase stops after the first 12 hours of culture; during the initial 12 hours of culture the cells are beginning to grow rapidly. Recent investigation has shown (WHITTAKER 1968) that synthesis of protein, as measured by L-valine-^{14}C incorporation into the acid-insoluble cell fraction, begins to increase rapidly during the second 6 hours of culture (6—12 hours) to a rate slightly less than twice the rate in early culture (0—6 hours). Also ribonucleic acid synthesis, assayed by uridine-^{3}H incorporation, begins to increase even faster, reaching a level during the second 6 hours of almost three times the initial rate. After 12 hours of culture the rates of both protein and RNA synthesis continue to rise. The correlation of increased rates of protein and RNA synthesis with lowered tyrosinase synthesis may, as suggested previously (WHITTAKER 1965), be related to some competitive effect operating at the site of protein synthesis.

Investigations by the HOLTZER group of the loss of phenotypic characters in cultures of cartilage and muscle cells have suggested to these workers that the initiation of new DNA synthesis is a triggering mechanism which de-activates certain genes associated with differentiation, thereby explaining the apparent antagonism between growth and differentiation (STOCKDALE and HOLTZER 1961; ABBOTT and HOLTZER 1966). Recently CAHN and LASHER (1967) have shown, in apparent contradiction of the results of ABBOTT and HOLTZER (1966) and NAMEROFF and HOLTZER (1967), that hyaluronidase-sensitive mucopolysaccharide matrix (chondroitin sulfate) is synthesized simultaneously with DNA in clonal cultures of chick embryo chondrocytes. This chondroitin sulfate synthesis might be caused by continued enzyme synthesis from previously formed messenger RNA or even from persistent stable enzymes, but the high percentage (20—30%) of cells undergoing this dual synthesis and the fact that clonal cultures of chondrocytes do retain their phenotypic characters (COON 1966) argues strongly against the DNA synthesis — dedifferentiation correlation observed by the Holtzer group being a causal relationship.

Examination of DNA synthesis in cultured pigment cells, using thymidine-^{3}H incorporation into the cold acid-insoluble cell fraction, shows that during the first 12 hours of culture virtually no DNA synthesis occurs (WHITTAKER 1968). DNA synthesis does not begin until during the second 12 hours of culture, and most tyrosinase synthesis has stopped prior to the surge in DNA synthesis. To explore this possible relationship still further, the effect of mitomycin C (an inhibitor of DNA synthesis) was tested on pigment cells. The lowest level of mitomycin C (5 μg/ml) which still caused maximum inhibition of DNA synthesis was found to have no effect of retarding the loss of melanin synthesis during the initial period of DNA synthesis (12—36 hours of culture). Based on deoxyribonuclease tests, this maximum level of inhibition caused by mitomycin C apparently represents total inhibition of DNA synthesis. Diminished tyrosinase in pigment cell cultures is probably not caused by the initiation of DNA synthesis.

An explanation of cell modulation as gene repression and derepression is almost irresistible at a time when intensive genetic analysis of cell function is being pursued vigorously and successfully. The experiments of DAVIDSON, EPHRUSSI and YAMAMOTO (1966) in which melanin-producing hamster melanoma cells were hybridized with non-pigmented mouse cells show that the genome of unpigmented cells represses the expression of tyrosinase genes. Melanotic expression is presumably under negative control in normal pigment cells as well. Perhaps all changes in tyrosinase production, including cell culture modulations, operate through the same mechanism. An alternative is suggested in the following section.

D. Messenger Competition and Stability

The relatively simple modulation phenomenon of pigment cell cultures could be explained by a plausible mechanism that does not require specific gene repressions and derepressions. Differentiated embryonic pigment cells *in situ* are increasing in protoplasmic volume (but not in number) at $6^1/_2 - 7$ days of incubation (COULOMBRE 1955); the *in vitro* culture conditons cause additional growth as well as cell division, and one of their responses is a rapid increase in the rate of synthesis of new RNA. I propose that this has the effect of flooding the protein-synthesizing (translational) mechanism with a surplus of new messenger RNA. If tyrosinase genes continue to be transcribed at their previous rate, production of the enzyme would diminish through competition for translation.

If the loss of tyrosinase synthesis is due simply to translation-level competition rather than repression of gene transcription, then several reasonable predictions can be made. A competition mechanism might allow at least a small amount of enzyme synthesis; production would not necessarily stop completely. A low level of synthesis does in fact occur. Also the loss of synthetic ability should be readily reversible with altered growth conditions, which it is indeed. A third prediction is that drastic interference with new RNA synthesis in culture would have the effect of stimulating tyrosinase synthesis at a time when it is normally declining, assuming that messenger RNA for tyrosinase is moderately stable.

A level of actinomycin D ($3 \mu g/ml$) which inhibits essentially all of new RNA synthesis in cultured pigment cells *does* have a stimulatory effect on both melanin biosynthesis and tyrosinase activity when pigment cells are treated with it during their first 12 hours in culture (WHITTAKER 1965, 1968). Actinomycin D-treated cells have higher rates of tyrosine incorporation into melanin and also a higher level of tyrosinase activity in homogenates of treated cells. Since the increase in level of tyrosinase is sensitive to cycloheximide inhibition, this rise is apparently due to new synthesis of the enzyme rather than to changes in the tyrosinase decay rate or to an enzyme activation phenomenon. Interestingly, cultures of $24-36$ hours of age are completely refractory to the stimulating effect of actinomycin D. This is to be expected since the surge of new RNA synthesis has occurred already.

Tests of messenger stability, beginning with 24-hour old cultures of pigment cells and measuring the decay of the rate of protein synthesis (incorporation of L-leucine-^{14}C) in the presence of actinomycin D ($3 \mu g/ml$), show that the rate of protein synthesis decays exponentially with a half-life of approximately 12 hours (WHITTAKER 1968). This suggested high stability of messenger RNA would contribute to the competition effect and also would exert a considerable damping effect on the recovery of enzyme synthesis following abrupt growth changes. An experiment to examine the stability of tyrosinase messenger RNA suggests that it is also reasonably stable. Cells which are inhibited for 24 hours with actinomycin D have a much higher rate of melanin biosynthesis at $18-24$ hours than they do at $12-18$ hours, thereby showing that tyrosinase messenger RNA is still active after 18 hours of no competition from new messenger RNA. Since actinomycin D does not raise the level of enzyme measured in older cultures we know that it does not retard the decay rate of the enzyme.

Admittedly, most of the results discussed could be as easily explained by a gene repression hypothesis. The actinomycin D might prevent synthesis of the repressor and fail to have an effect on the older cultures because the repressor had been made already. Also, if repression is not total, a low level of enzyme could occur in older cells. The one major point that is difficult to explain in this was is why, if the tyrosinase messenger RNA is relatively stable, most enzyme synthesis stops so soon. Competition with new messenger RNA seems to provide a more reasonable explanation of this fact.

IV. Summary and Conclusions

Investigation of the changes involved in loss of melanotic phenotype in pigment cell cultures provides a coherent description of what actually occurs in the cells: (1) cell pigmentation and pigment synthesizing components are diluted by growth, (2) the essential enzyme, tyrosinase, undergoes decay of activity, and (3) synthesis of tyrosinase virtually disappears. All three occur within a relatively short time, that is, a few hours.

The most important question is what regulates synthesis of tyrosinase. Considering the various facts that are presently known about pigment cell behavior, I favor the theory that enzyme regulation in *this* system is a simple competitive response at the level of messenger RNA translation in the cell, whereby tyrosinase synthesis diminishes during rapid growth because its messenger RNA is outcompeted for translation. According to this theory, any conditions that disrupted RNA synthesis might shift the balance in favor of higher tyrosinase synthesis and maintenance of phenotype. This conservative hypothesis has at least the charm of simplicity and is testable in a number of ways.

If this hypothesis is true for pigment cells, an important question is how applicable it might be to other examples of phenotypic instability in culture. There is a great diversity in the responses of different cell types to culture conditions (DAVIDSON 1964; DEFENDI 1964) and, at best, this competition may be only part of what takes place in other kinds of cells. A much larger problem is whether differentiation activities of cells in culture correspond in any way to the behavior of cells in the organism. Possibly the loss of phenotype in certain neoplastic cells can be explained adequately by the kinds of changes outlined in this review. Melanoma cells especially, seem capable of reversible losses of phenotype in the organism (GRAY and PIERCE 1964) which may be similar to those observed in vitro (MOORE 1964). On the other hand, a closely studied example of pigment cell change during Wolffian lens regeneration from iris epithelium in the newt (EGUCHI 1963; KARASAKI 1964) suggests that losses of cell phenotype in regeneration are too complex to be resolved by any simple explanation.

References

ABBOTT, J., and H. HOLTZER: The loss of phenotypic traits by differentiated cells. III. The reversible behavior of chondrocytes in primary cultures. J. Cell Biol. **28**, 473—487 (1966).

CAHN, R. D., and M. B. CAHN: Heritability of cellular differentiation: Clonal growth and expression of differentiation in retinal pigment cells *in vitro*. Proc. nat. Acad. Sci. (Wash.) **55**, 106—114 (1966).

—, and R. LASHER: Simultaneous synthesis of DNA and specialized cellular products by differentiating cartilage cells *in vitro*. Proc. nat. Acad. Sci. (Wash.) **58**, 1131—1138 (1967).

CHIAN, L. T. Y., and G. F. WILGRAM: Tyrosinase inhibition: its role in suntanning and in albinism. Science **155**, 198—200 (1967).

COON, H. G.: Clonal stability and phenotypic expression of chick cartilage cells *in vitro*. Proc. nat. Acad. Sci. (Wash.) **55**, 66—73 (1966).

—, and R. D. CAHN: Differentiation *in vitro*: effects of Sephadex fractions of chick embryo extract. Science **153**, 1116—1119 (1966).

COULOMBRE, A. J.: Correlations of structure and biochemical changes in the developing retina of the chick. Amer. J. Anat. **96**, 153—190 (1955).

Daniel, C. W., and K. B. de Ome: Growth of mammary glands *in vivo* after monolayer culture. Science **149**, 634—636 (1965).

Davidson, E. H.: Differentiation in monolayer tissue culture cells. Advanc. Genet. **12**, 143—280 (1964).

Davidson, R. L., B. Ephrussi, and K. Yamamoto: Regulation of pigment synthesis in mammalian cells, as studied by somatic hybridization. Proc. nat. Acad. Sci. (Wash.) **56**, 1437—1440 (1966).

Defendi, V. (Ed.): Retention of functional differentiation in cultured cells, pp. 1—116. Philadelphia: Wistar Institute Press 1965.

Eguchi, G.: Electron microscopic studies on lens regeneration. I. Mechanisms of depigmentation of the iris. Embryologia **8**, 45—62 (1963).

Grant, P.: Informational molecules and embryonic development. In: The biochemistry of animal development, Vol. I, pp. 493—494 (R. Weber, ed.). New York: Academic Press 1965.

Gray, J. M., and G. B. Pierce Jr.: Relationship between growth rate and differentiation of melanoma *in vivo*. J. nat. Cancer Inst. **32**, 1201—1210 (1964).

Hilfer, S. R.: The stability of embryonic chick thyroid cells "*in vitro*" as judged by morphological and physiological criteria. Develop. Biol. **4**, 1—21 (1962).

Karasaki, S.: An electron microscopic study of Wolffian lens regeneration in the adult newt. J. Ultrastruct. Res. **11**, 246—273 (1964).

Moore, G. E.: *In vitro* cultures of a pigmented hamster melanoma cell line. Exp. Cell Res. **36**, 422—423 (1964).

Moyer, F. H.: Genetic variations in the fine structure and ontogeny of mouse melanin granules. Amer. Zool. **6**, 43—66 (1966).

Nameroff, M., and H. Holtzer: The loss of phenotypic traits by differentiated cells. IV. Changes in polysaccharides produced by dividing chondrocytes. Develop. Biol. **16**, 250—281 (1967).

Priest, R. E., and J. H. Priest: Redifferentiation of connective tissue cells in serial culture. Science **145**, 1053—1054 (1964).

Sato, G., L. Zaroff, and S. E. Mills: Tissue culture populations and their relation to the tissue of origin. Proc. nat. Acad. Sci. (Wash.) **46**, 963—972 (1960).

Stockdale, F., and H. Holtzer: DNA synthesis and myogenesis. Exp. Cell Res. **24**, 508—520 (1961).

Twarog, J. M., and B. L. Larson: Induced enzymatic changes in lactose synthesis and associated pathways of bovine mammary cell cultures. Exp. Cell Res. **34**, 88—99 (1964).

Weiss, P.: Differential growth. In: The chemistry and physiology of growth pp. 135—186. (A. K. Parpart, ed.) Princeton: Princeton University Press 1949.

Whittaker, J. R.: Changes in melanogenesis during the dedifferentiation of chick retinal pigment cells in cell culture. Develop. Biol. **8**, 99—127 (1963).

— A relationship between increased protein synthesis and loss of melanin synthesis in monolayer cultures of chick retinal pigment cells. Amer. Zool. **5**, 643 (1965).

— An analysis of melanogenesis in differentiating pigment cells of ascidian embryos. Develop. Biol. **14**, 1—39 (1966).

— Loss of melanotic phenotype *in vitro* by differentiated retinal pigment cells: demonstration of mechanisms involved. Develop. Biol. **15**, 553—574 (1967).

— Translational competition as a possible basis of modulation in retinal pigment cell cultures. Submitted for publication (1968).

Clonal Aspects of Muscle Development and the Stability of the Differentiated State

Stephen D. Hauschka

Department of Biochemistry, University of Washington, Seattle, Washington 98102

I. Introduction

The stability of the differentiated state has the kind of catchy sonority which leaves the impression that some natural law is being stated; when, in fact, the phrase more truthfully refers to a broad, ill-defined, and unresolved set of questions with different connotations for different investigators. The various connotations are, in turn, intimately linked to numerous and diverse experimental systems, which can be considered under several broad operational categories: (1) the ability of differentiated cells *to retain* their specialized characteristics when cultured *in vitro*; (2) the ability of relatively undifferentiated cells *to retain* their capacity for differentiation after an extended proliferative or resting phase (as is the case for hemopoeitic stem cells); (3) the ability of cells *to leave* the differentiated state, undergo a proliferative phase, and then *return* to a differentiated state (as in the case of liver and amphibian limb regeneration); (4) the ability of nuclei from differentiated cells to function in foreign cytoplasmic environments.

With the advent of increasingly sensitive cytological and biochemical techniques, unambiguous experiments pertaining to these problems are now possible. A brief examination of several of these approaches will hopefully provide a broader background against which to judge the contribution which studies of skeletal muscle differentiation may eventually make to the problem as a whole.

The major tactical and theoretical obstacles associated with these systems are well illustrated by the current studies of cartilage differentiation. The bulk of experimental attention has been focused upon the question of a mutual exclusivity between differentiation and cell division.

When cultured in compact aggregates, chondrocytes retain their ability to synthesize the same sulfated mucopolysaccharides made *in vivo*. However, when dispersed and cultured as monolayers, chondrocytes commence DNA synthesis and lose their capacity to synthesize sulfated mucopolysaccharides; yet when again returned to their original compact environment, the proliferating chondrocytes cease DNA synthesis and recommence synthesis of their specialized product (Stockdale et al. 1963). More recent investigations in the same laboratory have shown that chondrocytes cultured as monolayers do not lose the capacity for synthesizing mucopolysaccharides *per se*, but rather their capacity for sulfating the mucopolysaccharide molecules (Nameroff and Holtzer 1967). However, it is not yet clear whether the species of mucopolysaccharide synthesized concomitantly with cell division is identical to the species which is sulfated under non-proliferating conditions. Interpretation is further complicated by the persisting synthesis of this mucopolysaccharide even after the sulfating capacity of the monolayer chondrocyte cultures has been reinstated due to cell crowding.

Somewhat in contrast to these findings, Cahn and Lasher (1967) have shown that 20—30 percent of the cells contained in colonies originating from chick embryo chondrocytes synthesize DNA concurrently with the incorporation of ^{35}S into a hyaluronidase-sensitive mucopolysaccharide. Unfortunately, the different culture conditions and methods of identifying sulfated mucopolysaccharides used by the two groups probably preclude a direct comparison of results; yet the extensive difference between their findings, if nothing else, underscores the effect of environmental conditions on the expression of phenotypic traits, and argues for some greater effort toward standardization of culture techniques.

The foregoing studies refer primarily to the capacity of chondrocytes to synthesize a limited set of specialized products. They say little about the genetic capability of the cells. This problem has been most extensively pursued by Coon (1966), who has demonstrated that the capacity of chondrocytes to form differentiated cartilage remains unimpaired after more than thirty generations *in vitro*. Operationally then, one is faced with the fact that cells can be caused to lose a phenotypic trait while retaining the genetic capacity to re-express the trait under more favorable environmental conditions. What one would like to know, is whether the loss of phenotypic expression stems from interactions at the genetic level; or, whether it results from interactions at the physiological level such as direct enzymatic regulation or altered permeability of the cell membrane.

Another approach to these problems has been through investigation of what might be called "the critical division hypothesis". Though couched in varying contexts, this hypothesis has been enunciated by numerous investigators. In essence, it proposes that cytodifferentiation results from an irreversible division during which the cell is influenced (most likely by the environment) to differentiate rather then continue dividing. This hypothesis has been tested by providing cells with the environmental requirements for their differentiation while, at the same time, eliminating their proliferative capacity with mitotic inhibitors.

Applying such techniques to explanted mammary glands, Stockdale and Topper (1966) have demonstrated that cell division is required for both morphological and chemical aspects of this tissue's differentiation. Furthermore, they have determined in which portions of the cell cycle the three hormones insulin, prolactin and hydrocortisone act to promote casein biosynthesis (Lockwood, Stockdale, and Topper 1967). Although these findings do not yet indicate whether the withdrawal from cell division regulates casein biosynthesis in a mechanistic sense, at present, they represent one of the most thorough approaches to the proliferation-differentiation question and they will undoubtedly shed further light on the subject in the future.

In all likelihood though, the exact relationship between cell division and cytodifferentiation will vary somewhat from tissue to tissue. Elucidation of the exact relationship may be particularly difficult in those tissues which are subjected to inductive influences relatively early in their developmental history. It may be hoped, however, that a few basic mechanistic similarities will apply throughout the various tissue types.

II. Muscle Development *in vivo**

A. Cytological Observations

The differentiation of skeletal muscle occurs over a comparatively long period of embryonic and even post-embryonic development. The first definitive muscle structures appear in the somites, where an anterior-posterior gradient of muscle differentiation follows the posterior progression of somite formation. At the same time, a dorsal-ventral gradient of myogenesis also appears, which spreads ventrally from the somites through the lateral mesoderm and finally into the limb musculature. The apparent progression of myogenesis from the somites into other areas led to the assumption that lateral skeletal muscle was of somite origin. However, this assumption was disproved by STRAUS and RAWLES (1953), who demonstrated that the lateral musculature develops directly from the lateral plate mesoderm.

The sequence of somite muscle cytodifferentiation has been described at the fine structure level by ALLEN and PEPE (1965), DESSOUKY and HIBBS (1965), PRZYBYLSKI and BLUMBERG (1966), OBINATA, YAMAMOTO, and MARUYAMA (1966), FISCHMAN (1967), and FIRKET (1967). The approach taken by ALLEN and PEPE is particularly impressive in that it combines histological observations with the isolation of structurally recognizable, negatively stained thick and thin filaments of myosin and actin (HUXLEY 1963; HANSON and LOWY 1963). By this device, the sometimes questionable identification of myosin and actin filaments seen in electron micrographs is measurably overcome.

Fine structure analysis of myotomal cells from the somites of stage 14 embryos (50 hours of incubation) disclosed no differences between these cells and so-called undifferentiated mesenchymal cells; nor was it possible to identify thick or thin filaments in negatively stained homogenates of stage 14 tissue. However, by stage 16 (53 hours of development) the majority of myotomal cells have become elongated; their mitochondria also assume a more elongated appearance and become oriented with the cell's long axis; and the cytoplasm becomes noticeably granular as small aggregates of ribosomes begin forming. At the same time, the first actin-like (55 to 65 Å diameter) thin filaments appear in the cytoplasm, and actin filaments are first detected in the negatively stained homogenates. After stage 16, the number of thin filaments continues to increase; but the thick filaments do not make their first appearance until stage 18 (65 hours of development). Interestingly, the thick filaments are never observed singly, but only in combination with thin filaments. It seems probable then, that the first thick filaments are generated within a hexagonal array of thin filaments, thus immediately forming the hexagonal lattice structure encountered in differentiated muscle (HUXLEY 1957). Coupled with the appearance of thick filaments, or perhaps preceding it slightly, large polyribosomes appear, some of which contain up to 75 individual ribosomes. Recent studies in which polysomes containing 50−60 ribosomes were identified as those synthesizing myosin (HEYWOOD, DOWBEN, and RICH 1967), suggest that the large ribosomal aggregates described in the histological studies are very likely those responsible for myosin

---

* The vast majority of descriptive and experimental studies on muscle development have been done with the chick embryo. Chicken muscle has also been the material of choice for almost all of the *in vitro* work. For this reason, the following description is based solely upon the chick literature; but it undoubtedly applies to muscle differentiation in most other species as well.

synthesis. Allen and Pepe (1965) also observed the first appearance of glycogen granules at about 65 hours of development.

Subsequent stages of somite muscle differentiation entail a continual increase in the number and size of actin-myosin aggregates, but the typical banding pattern of mature myofibrils does not appear until $5^1/_2$ days of development; and even after 4 days incubation, cells exhibiting all of the earlier characteristics of muscle cyto-differentiation are still present. Thus the differentiation of muscle cells within even so small a volume as a single myotome is highly asynchronous. The same asynchronous cytodifferentiation has also been observed with respect to the binding of fluorescent antimyosin (Holtzer, Marshall, and Finck 1957). These investigators report that the first positive detection of myosin in the most anterior cervical mytomes occurs at stage 13; thus the temporal difference in myosin appearance along the anterior-posterior axis is about 12 hours between the most anterior myotomes and the twentieth pair. An even earlier report of myosin in segmented somites has been described by Deuchar (1960), who selectively detected myosin at stage 11 (40 to 45 hours incubation) by its calcium-activated adenosine triphosphatase (ATPase) activity, and by its insolubility in 50 percent glycerol. Presumably, the myosin detected at this early stage was not aggregated, or thick filaments probably would have been recognized in the numerous electronmicroscopic studies of somite muscle development. If the calcium-activated ATPase is genuine myosin, then the apparent sequence of actin and myosin appearance described by Allen and Pepe (1965) may require further investigation.

The precise nature of the histological changes accompanying muscle differentiation in the limb has not been so thoroughly documented. However, the light microscopic observations of Kitiyakara (1959) have described an asynchrony with respect to the cytodifferentiation of individual cells similar to that observed in the development of myotomal muscle by Allen and Pepe (1965). Definitive limb buds are first visible at about 54 hours (the time at which actin filaments are already detectable in the myotomes); yet multinucleated muscle cells are not evident until the sixth day of development, and true muscle fibers do not appear in the limb until the seventh and eighth days. A sequential study of limb skeletal muscle formation at the fine structure level has not been reported; but the careful description of material from 12-day embryos suggests that the process is essentially similar to observations made on developing somitic muscle (Fischman 1967).

B. Biochemical Observations

A systematic and internally comparable set of measurements describing the chemical differentiation of muscle tissue has not yet been made by any single group of investigators; however, a large number of biochemical changes associated with muscle differentiation have been reported, and these are illustrated in Fig. 1. Due to the many different standardization procedures used in compiling these measurements, the relative quantities of various molecules are not directly comparable; yet the rates of change during development are probably safe to compare.

Even though histological observations of the limb region indicate the presence of definitive muscle fibers by the seventh day of development (Kitiyakara 1959), accurate chemical measurements of myosin have not yet been feasible until the ninth day (BARIL and HERRMANN 1967). At that time the myosin concentration is 0.67 mg per gram wet weight muscle (roughly one percent of the total protein); and by the eighth day post-hatching myosin accounts for nearly 10 percent of the total muscle tissue protein.

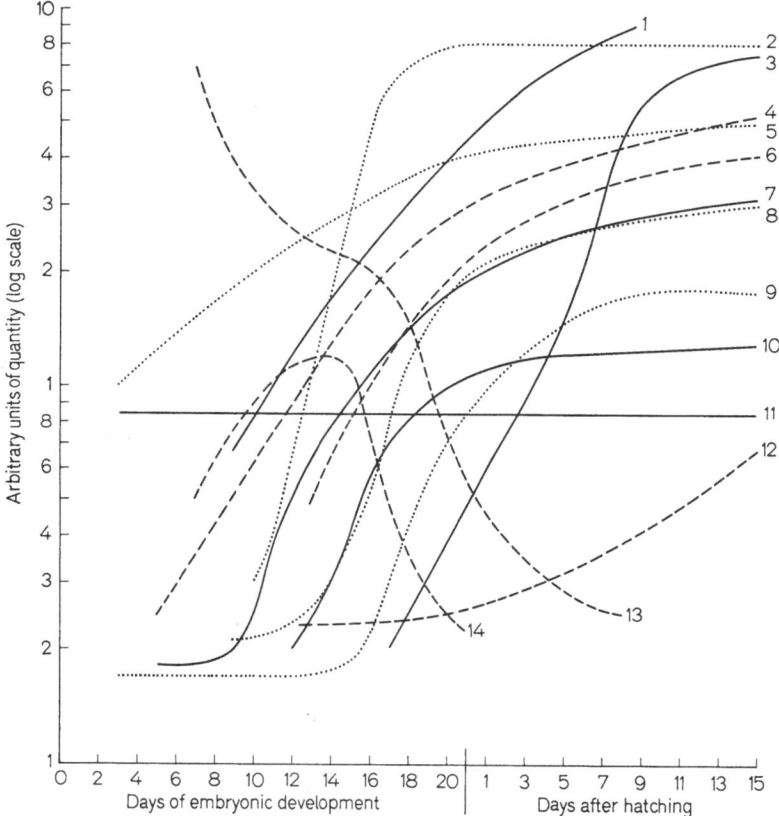

Fig. 1. A compilation of various biochemical measurements made on developing chick skeletal muscle (1 Myosin, 2 Collagen, 3 Phosphorylase, 4 Creatine kinase, 5 Aldolase, 6 Phospho-glycerokinase, 7 Glycogen, 8 Actomyosin, 9 Glutamate-oxalacetate Transaminase, 10 Adenylpyrophosphatase, 11 Isocitrate Dehydrogenase — Hexokinase — Phosphofructo-kinase, 12 Alkaline Phosphatase, 13 Percent nuclei synthesizing DNA, 14 Acetylcholin-esterase). The relative total concentrations are meaningless due to differences in the stan-dardization of data; but the rates of change during development may be safely compared if the possible effects of the various standardization procedures are kept in mind. The data were originally expressed in the following units, and have been replotted from the following sources: Myosin, mg chromatographically pure/gm muscle tissue (BARIL and HERRMANN 1967); Collagen, percent wet weight muscle (HERRMANN and BARRY 1955); Phosphorylase, activity/mg wet weight muscle; Creatine kinase, Aldolase, Phosphoglycerokinase, Gluta-mate-oxalacetate transaminase, Isocitrate dehydrogenase, Hexokinase, and Phosphofructo-kinase all in enzyme activity/mg wet weight muscle (EPPENBERGER et al. 1962/63); Glycogen, nmoles/gm wet weight muscle (ARESE, RINAUDO, and BOSIA 1967); Actomysin, mg/gm wet weight muscle (HERRMANN 1952); Adenylpyrophosphatase, activity/100 mg wet weight muscle (MOOG 1947); Alkaline phosphatase, activity/μg DNA (KONIGSBERG and HERRMANN 1955); Percent nuclei synthesizing DNA, autoradiographically determined (MARCHOK and HERRMANN 1967)

With the exception of aldolase and phosphoglycerokinase, most of the glycolytic and citric acid cycle enzymes measured show no significant changes during development (EPPENBERGER et al. 1962—1963). The same investigators also report though, that glyceraldehyde-3-phosphate dehydrogenase and pyruvate kinase are maintained at constant levels until hatching, at which time the enzyme concentrations drop precipitously to less than 50 percent of their day 19 values. Following hatching, the concentrations gradually rise, and attain the previous steady-state levels after about three weeks. No explanation for the anomalous behavior of glyceraldehyde-3-phosphate dehydrogenase and pyruvate kinase is yet apparent; neither is it known why aldolase and phosphoglycerokinase lag behind the other glycolytic enzymes in attaining their adult steady-state levels. It should be particularly instructive to reexamine the accumulation of these enzymes during embryonic development with reference to the "constant proportion" groups of Embden-Meyerhof and mitochondrial enzymes described by BÜCHER and coworkers (PETTE, LUH, and BÜCHER 1962; PETTE, KLINGENBERG, and BÜCHER 1962) to determine whether this phase of regulation exists throughout muscle development, or whether it appears secondarily — perhaps concomitantly with the appearance of multinucleated fibers.

The behavior of lactate dehydrogenase (LDH) represents another enigma. EPPENBERGER et al. (1962—1963) found no change in the total activity of LDH per gram wet weight skeletal muscle between the third day of embryonic development and adulthood. In contrast, KAPLAN and CAHN (1962) measured a change of greater than three orders of magnitude in breast muscle LDH activity over the same developmental time span. The latter authors also reported a sizable variation in LDH activity between different striated muscles, ranging from 80 units LDH per gram weight in the deltoid to 4,300 units in the inner breast muscle (pectoralis superficialis). The same magnitude of variation was also reported for isozymic types of LDH. Inner breast muscle contains less than one percent heart type LDH while the outer back muscles (latissimus dorsi and pars anterior) contain more than 99 percent heart type LDH (KAPLAN and CAHN 1962). These remarkable muscle-specific differences in LDH type and quantity suggest rather strongly that gross biochemical data (at least with respect to metabolic enzymes) may reflect an overall averaging of enzyme concentrations which bear only slight resemblances to true muscle-specific concentrations. An *in vitro* analysis of the behavior of LDH during the development of these various muscles may prove quite interesting, particularly since the environmental condition can be standardized.

Perhaps one of the most disconcerting aspects of the biochemical studies completed to date, is that the changes measured are not easily correlated with the more obvious physiological or histological changes accompanying skeletal muscle development. For example, five molecular species (collagen, adenylpyrophosphatase, isocitrate dehydrogenase, hexokinase, and phosphofructokinase) attain their steady-state concentrations prior to hatching, while the remainder of the molecular species measured (with the exception of acetylcholinesterase) continue to increase after hatching. Even the enzymes which appear late in development (glutamate-oxalacetate transaminase and phosphorylase), begin their rise in activity several days prior to hatching. Likewise, there is not yet any obvious correlation between the increase in multinuclearity of muscle fibers (shown inversely in Fig. 1 as the percent nuclei synthesizing DNA) and the acquisition of differentiated biochemical functions. Even among the other relatively tissue-specific proteins measured (muscle phosphorylase, creatine kinase, and acetylcholinesterase), none exhibit profiles noticeably similar to the myosin or actomyosin curves with respect to the timing of accumulation.

To a certain extent, however, these results were to be expected since anatomical observations have long indicated that the development of skeletal muscle continues well beyond embryonic life. But, other significant contributions to the overall smearing out of the biochemical changes which accompany muscle differentiation are undoubtedly the type of muscle-specific variations reported by KAPLAN and CAHN (1962); and perhaps even more significant, the change in cell population types within a given muscle — particularly the change in fibroblast to myoblast ratio.

III. Muscle Differentiation *in vitro*

A. Mixed Population Cultures

Although explants of skeletal muscle tissue were grown *in vitro* as long as fifty years ago (LEWIS and LEWIS 1917), studies of monodisperse embryonic muscle cell cultures originated less than ten years ago (KONIGSBERG 1960). The triumph of these latter studies was largely that the differentiation of an embryonic tissue could at long last be followed *in vitro*; and when viewed against the fatalistic tendency toward "dedifferentiation" which was then common for cultured cells, the triumph was an exciting one.

Perhaps the most significant observations to follow from these studies were analyses of how muscle fiber multinuclearity originates. Based upon their measurements of DNA content per nucleus in regenerating mouse muscle, LASH, HOLTZER, and SWIFT (1957) had already presented strongly suggestive evidence that multinucleated muscle fibers arise through cell fusion. These investigators had shown that nuclei contained in muscle fibers have exclusively 2n DNA values, while the nuclei in surrounding cells (which were actively dividing) have DNA values distributed between 2n and 4n. Other indirect experiments suggesting that multinuclearity did not originate through amitotic nuclear division, were based upon the finding that inhibition of DNA synthesis by nitrogen mustard did not block the further development of muscle syncytia (KONIGSBERG et al. 1960). Soon afterwards, direct support for the role of cellular fusion in the formation of multinucleated muscle fibers was obtained by time-lapse cinematography (CAPERS 1960; COOPER and KONIGSBERG 1961). An even more convincing demonstration of the role of cell fusion in syncytial muscle development has been provided by YAFFE and FELDMAN (1965), who described the formation of hybrid multinucleated muscle fibers in mixed cultures of embryonic rat, rabbit, calf, and chicken myoblasts. Recent biochemical support for the fusion of myoblasts has resulted from studies of electrophoretically variant enzymes in allophenic mice. Analysis of skeletal muscle isocitrate dehydrogenase indicated the presence of a hybrid enzyme type as well as both donor types; whereas electrophoresis of liver homogenates from the allophenic mice disclosed only the two donor enzyme variants. Presumably then, both isocitrate dehydrogenase genes were functioning in muscle fibers which resulted from the fusion of myoblasts derived from each parental cell line, and the polypeptide products were able to form hybrid enzyme molecules within the common cytoplasm (MINTZ and BAKER 1967). Electron microscopic analysis of muscle differentiation *in vivo* has also supported the principle of cell fusion (HAY 1963; SHAFIG 1963) and suggests that fusion occurs through vesiculation of the opposed cellular membranes.

In this context, the reports of FIRKET (1958), and OKAZAKI and HOLTZER (1966) that cell fusion occurs non randomly with respect to the cell cycle are particularly intriguing. Indeed, they discovered that myoblasts did not enter muscle fibers until a minimum of 8 hours had elapsed following the completion of DNA synthesis, as if to suggest that the mitotic cycle was accompanied by cyclical fluctuations in membrane characteristics which determine the ability of cells to fuse. HOLTZER's group has also investigated the relationship of mitosis to the synthesis of muscle-specific proteins, and have found that DNA and myosin synthesis appear to be mutually exclusive phenomena in somite muscle cells — at least at the level of detection afforded by fluorescent microscopy (STOCKDALE and HOLTZER 1961). These findings have recently been corroborated for leg skeletal muscle by COLEMAN, COLEMAN, and ROY (1966).

Although high population monodisperse cell cultures do not provide a clean separation of fibroblasts and myoblasts, they do offer some advantages over *in vivo* systems for biochemical studies of muscle differentiation — particularly with regard to the initiation of cell fusion (i. e. a comparatively large number of myoblasts begin forming multinucleated fibers at about the same time). As a consequence, the correlation of biochemical changes with the onset of multinuclearity, which is distributed over such an extended period *in vivo*, can be examined much more advantageously *in vitro*.

COOPER and KONIGSBERG (1961) have qualitatively assessed the concentration of the mitochondrial enzyme succinate dehydrogenase histochemically, and find that it increases markedly after cell fusion. Likewise, immuno-fluorescence assays of the soluble glycolytic enzyme, glyceraldehyde-3-phosphate dehydrogenase, detect an increase in concentration following fusion (EMMART, KOMINZ, and MIGUEL 1963); and the augmented activity of creatine kinase has also been correlated with an increase in multinucleatiry (REPORTER, KONIGSBERG, and STREHLER 1963). However, since the latter authors were unable to detect creatine kinase until the seventh day of culture, whereas a detectable enzyme level was already present in the 11-day chick embryo source tissue, it seems likely that the appearance of creatine kinase activity might be more meaningfully associated with some other differentiated activity than with an immediate effect of cell fusion (which undoubtedly commenced by at least the fourth day of culture).

Somewhat different and more generalized parameters of muscle differentiation have been studied by YAFFE and FUCHS (1967). Using autoradiographic techniques, these investigators have measured the relative incorporation of RNA and protein precursors by fused and unfused cells. They find, that whereas the incorporation of uridine into muscle fiber nuclei is 3 to 4 times less than into the nuclei of mononucleated cells, the incorporation of leucine is equivalent in mononucleated and multinucleated cells. One serious difficulty with their interpretation, though, is that it fails to distinguish between incorporation by myoblasts and fibroblasts. Since the bulk of comparative measurements were apparently made with relatively old cultures (i. e. after the vast majority of mononucleated myoblasts have already been eliminated by fusion), the comparison was actually being made between two different cell types: multinucleated muscle cells and mononucleated fibroblasts. The difference in uridine incorporation by the two populations may thus have nothing to do with multi-nuclearity *per se*, but may be attributed solely to intrinsic differences between fibroblasts and myoblasts.

B. Clonal Studies of Muscle Development

As has been repeatedly stressed, embryonic muscle tissue consists of at least two cell types: myoblasts and fibroblasts. Hence biochemical, and even cytological, investigations of myogenesis which neglect consideration of these cell type differences may err in their conclusions. A clonal analysis of myogenesis thus has two significant advantages over mass culture and *in vivo* studies: (1) it permits clean separation of the various cell types of which embryonic muscle tissue is composed, thus enabling biochemical studies to be directed at a uniform cell type; (2) it permits determination of the extent to which the various embryonic cell types depend upon one another for their subsequent differentiation.

One serious drawback to clonal studies, which should be kept in mind from the start, has been the severe limitation of cellular material available for biochemical experiments. A healthy 2-week muscle colony may contain in the order of 50,000 nuclei; thus a prohibitive number of such clones would be required to obtain one milliliter of packed cells. This limitation is largely responsible for the fact that no biochemical measurements have yet been published on clonal muscle cell populations. To a considerable extent, however, the situation may be remedied by resorting to sub-cloned populations (KONIGSBERG and HAUSCHKA 1965). By this procedure, an amplification of several thousand-fold may be achieved; thus when coupled with the use of biochemical micro techniques, and particularly the use of radioisotopes, the problem of sufficient material should be largely eliminated.

1. *Cellular Interaction in Muscle Development*

From the standpoint of studying interrelationships between the various embryonic cell types of which developing muscle is composed, clonal analysis has already proved extremely valuable. In fact, unlike many other embryonic tissues which consist of sharply demarcated epithelial and mesenchymal components, embryonic muscle consists of a rather thorough intermingling of myoblasts and fibroblasts. Single cell studies were thus the only solution to the mixed cell population problem. By applying this technique, we have shown that the *in vitro* differentiation of skeletal muscle depends upon an interaction between fibroblasts and myoblasts which is mediated by collagen (KONIGSBERG and HAUSCHKA 1965; HAUSCHKA and KONIGSBERG 1966).

The suspicion that a second embryonic cell type was necessary for muscle development derived from the finding that isolated myoblasts depend upon the presence of "conditioned medium" in order to form differentiated muscle colonies (KONIGSBERG 1963). Conditioned medium is a medium which has been exposed for three days to the metabolic activities of densely populated mixed cell cultures derived from embryonic leg muscle. When removed from these cultures and used for establishing clonal cultures, the conditioned medium is then able to support a highly significant increase in the percentage of colonies which differentiate as striated muscle. To investigate some of the biological parameters of the conditioned medium requirement we set out to determine when, during the development of a muscle colony, conditioned medium was required. This question seemed particularly interesting due to the bi-phasic nature of muscle colony differentiation (KONIGSBERG 1963).

The initial phase in the temporal development of muscle colonies consists of a 4—5 day period of rather rapid cell division, with at least some cells exhibiting doubling times of less than 14 hours (see Table 1). Mitotic asynchrony is remarkably evident in many colonies by the second cell division, and by the fourth and fifth days it

Table 1. *Variation in the growth rate of muscle colonies*

		Number of cells per colony							
	2	3—4	5—8	9—16	17—32	33—64	65—128	129—256	257—512
		Minimal number of exponential cell divisions required							
	1	2	3	4	5	6	7	8	9
		Percent colonies of given size							
Day 3	23.6	27.4	22.6	21.7	4.7	0	0	0	0
Day 4	5.4	12.5	15.5	15.5	23.8	25.5	1.8	0	0
Day 5	10.4	4.9	4.3	18.9	15.8	14.6	20.7	9.8	0.6

The data presented above are from a single experiment in which petri plates received 400 cells each from the same primary cell suspension. Five petri plates were fixed and stained each day after 3, 4 and 5 days of culture. Muscle colonies were recognized by the predominance of bipolar cells (cf. Konigsberg 1963). The number of colonies counted on days 3, 4, and 5 respectively were 106, 168, and 122

is pronounced. There is also a great variation in the growth rate between different muscle colonies as illustrated by the data in Table 1 (Hauschka 1966). Cell fusion usually begins by the fifth or sixth day; and from that time on, fusion and proliferation continue side by side in the same colony. On the basis of these observations, it seemed possible that the conditioned medium requirement might be restricted to a single phase — either the initial proliferative phase or the later fusion phase of muscle colony development.

To test these alternatives, cells were established in conditioned medium, and switched to unconditioned medium after various intervals. In some cases as little as one day's exposure to conditioned medium proved sufficient to support the normal high percentage of muscle colony differentiation characteristic of control cultures maintained in conditioned medium throughout the 2-week culture period. Our initial conclusion from this set of experiments was that conditioned medium was not required for the cell fusion phase of muscle colony differentiation; but the one day requirement for conditioned medium, in turn, seemed so short that it appeared doubtful whether myoblasts even required the presence of conditioned medium for proliferation. An alternative to rejection of both the former hypotheses, was that we had not properly tested the temporal requirement of myoblasts for conditioned medium. Certainly, we had demonstrated that myoblasts did not require perpetual contact with conditioned medium itself; but the brevity of the necessary exposure period suggested that perhaps some residue remaining after removal of the bulk of the conditioned medium was sufficient to satisfy the requirement.

This possibility was tested by pretreating petri plates with conditioned medium prior to the addition of cells. After three days of treatment the conditioned medium was removed. The plates were then rinsed several times with distilled water and with unconditioned medium. An aliquot of cells contained in unconditioned medium was then added to each petri plate. It was found that such pretreated plates, the cells on which had never been exposed directly to conditioned medium, supported the same percentage of muscle colony differentiation as occurred under conditions in which the cells had been cultured continuously in conditioned medium. Control cultures established in petri plates pretreated with unconditioned medium did not have the ability to support muscle colony development. It could thus be concluded that some component of conditioned medium was deposited upon the petri plate surface during the pretreatment period, and that the myoblasts reacted with this component during their subsequent differentiation. The special attribute of conditioned medium was then not due to supplementation of a simple nutritional deficiency in the medium formulation, but rather to the presence of a more unique component which might be thought of as altering the surface properties of the petri plate in a fashion suitable for muscle differentiation.

In order to determine the identity of such a component, we decided to reassess the cell and culture requirements necessary for making effective conditioned medium. This investigation turned up two interesting facts. First, the most effective conditioned medium was prepared from secondary cultures of fibroblast-like cells (KONIGSBERG and KUPRES 1965–1966). Second, the favored culture conditions for conditioned medium preparation were associated with the stationary phase of confluent fibroblast monolayers. Medium exposed to the metabolic activities of such stationary cell populations for three days was most effective.

2. *Molecular Mechanism of the Fibroblast-Myoblast Interaction*

When combined with our previous suspicion that the effective conditioned medium component acted by altering the petri plate surface, these were particularly intriguing clues. Even more so, because collagen, the most characteristic fibroblast product, is synthesized predominantly during the stationary phase and is deposited extracellularly (GREEN and GOLDBERG 1964). Electron microscopic examination of the cultures responsible for making conditioned medium substantiated the occurrence of periodic, collagenlike extracellular fibrils (KONIGSBERG and HAUSCHKA 1965). It thus seemed worthwhile to test the question whether or not collagen was the effective component of conditioned medium.

This was accomplished by purifying collagen from rat tail tendons, spreading a thin collagen film on the petri plate surface, and growing colonies upon the collagen substratum in the presence of unconditioned medium. The results of such an experiment with the necessary controls are shown in Figs. 2 and 3. These findings unambiguously demonstrated that collagen could replace the conditioned medium requirement (HAUSCHKA and KONIGSBERG 1966).

Since collagen replaces conditioned medium, and since the same cell type (the fibroblast) which makes effective conditioned medium is characterized by its ability to synthesize collagen, and is present, moreover, in developing muscle tissue, it seemed possible that *in vivo* muscle differentiation might be mediated through association of myoblasts with the collagen fibrils deposited by fibroblasts. However, the foregoing experiments in no way proved that collagen was, indeed, the effective component in conditioned medium. In order to extend the *in vitro* analogy to *in vivo* conditions, it was therefore crucial to determine whether conditioned medium contained newly synthesized collagen; and, if so, whether this collagen was bound to the surface of petri plates pretreated with conditioned medium. Supporting evidence in this regard was obtained through the preparation of conditioned medium in the presence of ^{14}C-proline.

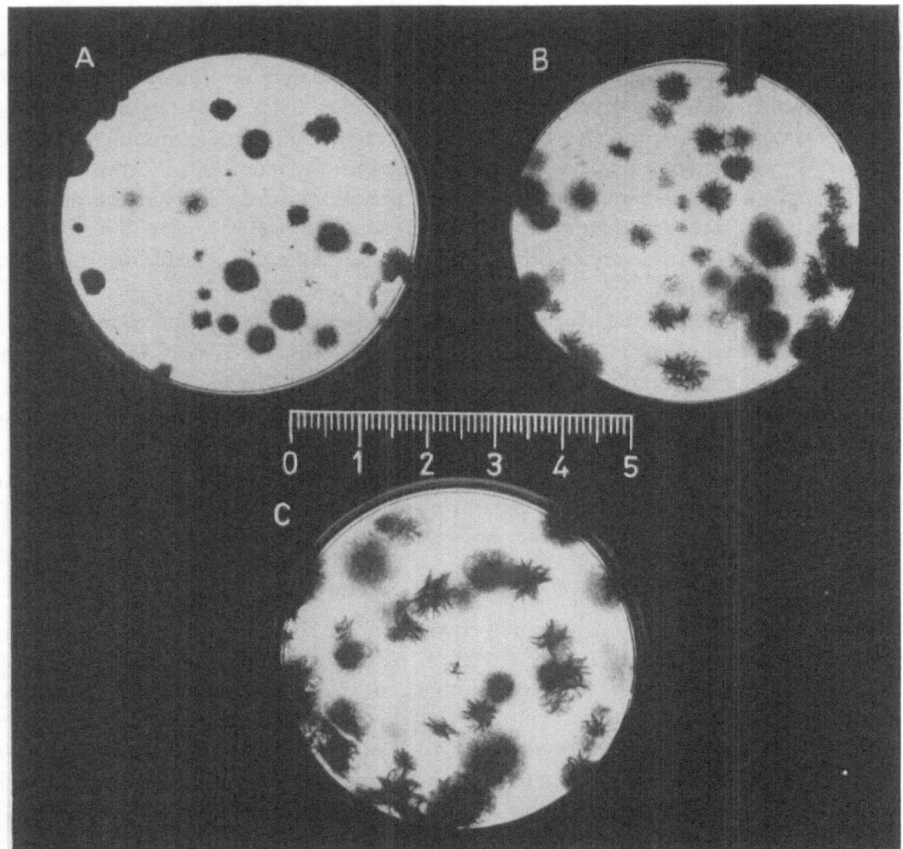

Fig. 2. Three representative Petri plates illustrating the ability of a collagen substratum to replace the requirement for conditioned medium in the development of muscle colonies. (A) Control culture: untreated surface; unconditioned medium. (B) Control culture: untreated surface; conditioned medium. (C) Colonies which have developed in unconditioned medium on a surface of reconstituted rat tail collagen. Each culture was inoculated with an equal aliquot (400 cells) from the same suspension of trypsin-dissociated embryonic (12 day) leg muscle cells. Cultures were fixed and stained on the 13th day of culture. (From Hauschka and Konigsberg 1966)

The identification of newly synthesized collagen is greatly facilitated by its unique chemical properties and mode of biosynthesis. Unlike all other vertebrate proteins (except elastin), collagen contains the imino acid hydroxyproline. Collagen is easily distinguished from elastin on the basis of its solubility, its sensitivity to collagenase, and its characteristic proline to hydroxyproline ratio. Using these criteria, the presence of soluble, newly synthesized collagen was demonstrated in conditioned medium, and it was shown that the newly synthesized collagen was strongly adsorbed to the surfaces of petri plates pretreated with the radioactive conditioned medium (Hauschka 1966). Thus the replacement of conditioned medium by a purified collagen substratum would seem to represent a truthful substitution of the precise molecule responsible for mediating the fibroblast-myoblast interaction.

Regarding the specificity of the collagen-myoblast interaction, it should be mentioned that adsorption of serum proteins to the petri plate surface does not alleviate the collagen dependence, and fibrin clots were likewise ineffective. However, gelatin,

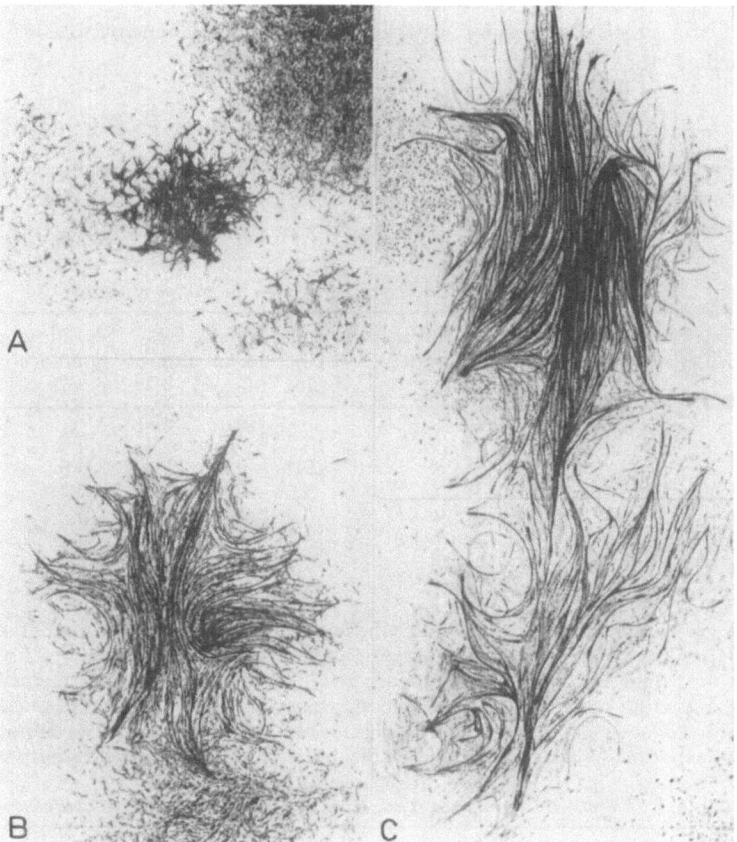

Fig. 3. Comparison of representative muscle colonies which develop under each of the three sets of conditions tested. (A) Small colony containing a few multinucleated muscle fibers grown on an untreated Petri plate in unconditioned medium. Those few muscle clones which develop under these conditions are most frequently observed in close proximity to colonies of fibroblast-like cells. It seems quite probable that collagen produced by these neighboring fibroblast-like colonies is sufficient to support myogenesis to a limited extent. (B) Muscle colony of a culture grown in conditioned medium; muscle fibers are considerably longer and more numerous than in (A). (C) Two muscle colonies typical of those which develop on a collagen-coated surface in unconditioned medium. Notice the overall fiber orientation which may be due to orientation of the supporting collagen substratum. (From HAUSCHKA and KONIGSBERG 1966)

a denatured form of collagen, was equally effective as collagen (KONIGSBERG and KUPRES 1965—1966). Thus the periodic fine structure of native collagen fibrils may not be essential to the collagen-myoblast interaction. Furthermore, no species specificity was exhibited with respect to the collagen source. Collagen and gelatin prepared from rat-tail tendon, carp swim bladder, cow hides, and chicken-foot tendons were equally effective (Hauschka, Konigsberg, and Kupres 1966), suggesting a certain lack of positional side chain specificity. Attempts to replace collagen with its acid hydrolysate, and with the smaller peptides released by collagenase digestion were all ineffective, thereby ruling out a simple nutritional mechanism. However, the precise molecular mechanism of the collagen-myoblast interaction awaits further elucidation.

IV. Collagen and the Stability of Myoblast Differentiation

The apparent dependence of muscle development upon collagen suggests a number of questions related to the stability of myoblast differentiation, and also the intriguing possibility that collagen may be used as a device to influence directly the sub-

Table 2. *The effect of a collagen substratum on plating efficiency*

Expt. No.	No. petri plates of each type	Mean plating efficiency Unconditioned medium	Uncond. medium plus collagen	Significance of the difference between means
1	6	11.0 ± 1.0	11.5 ± 0.5	>0.8
2	8	8.6 ± 0.4	9.1 ± 0.4	>0.6
3	6	7.9 ± 0.6	9.0 ± 0.6	>0.4
4	6	8.9 ± 0.4	8.4 ± 0.2	>0.5
5	5	6.7 ± 1.0	8.3 ± 1.0	>0.4
6	6	6.4 ± 0.5	7.7 ± 1.0	>0.5
7	10	10.3 ± 0.6	14.3 ± 0.5	>0.02

Values are expressed as mean plating efficiency [(total colonies per plate)/(no. of cells inoculated per plate) $\times$ 100] per Petri plate $\pm$ standard error of the mean. Probabilities have been determined according to Student's t-test. In none of the experiments were the differences in plating efficiency between unconditioned medium groups and unconditioned medium plus collagen groups statistically significant. Accompanying data from the same sequence of experiments are presented in Table 3. (From Hauschka and Konigsberg 1966)

Table 3. *The effect of a collagen substratum on the development of muscle clones*

Expt. No.	Unconditioned medium Mean no. of muscle clones	Mean no. of fibro clones	Mean percent muscle	Unconditioned medium plus collagen Mean no. of muscle clones	Mean no. of fibro clones	Mean percent muscle
1	1.2 ± 0.4	42.2 ± 3.3	2.6 ± 0.9	31.0 ± 1.1	15.2 ± 0.8	67.3 ± 2.5
2	0.13 ± 0.05	34.4 ± 1.2	0.4 ± 0.4	21.9 ± 1.4	14.8 ± 1.4	59.9 ± 3.9
3	0.2 ± 0.2	31.5 ± 2.4	0.6 ± 1.8	19.0 ± 0.9	17.0 ± 1.9	53.6 ± 0.9
4	0.0 ± 0.0	35.3 ± 2.3	0.0 ± 0.0	20.8 ± 1.3	12.7 ± 0.8	62.2 ± 2.3
5	0.0 ± 0.0	27.2 ± 6.7	0.0 ± 0.0	16.6 ± 1.1	17.6 ± 1.5	47.8 ± 5.5
6	1.5 ± 0.5	24.2 ± 1.5	6.4 ± 1.9	21.7 ± 0.7	10.3 ± 0.9	67.7 ± 2.3
7	0.2 ± 0.4	41.1 ± 1.9	0.5 ± 0.3	31.9 ± 1.6	25.4 ± 1.3	55.7 ± 1.9

Values are expressed as mean number colonies per Petri plate $\pm$ standard error of the mean. Mean per cent muscle value [(muscle colonies)/(total colonies) $\times$ 100] are presented since they normalize the data with respect to fluctuations of plating efficiency from experiment to experiment. Tests of the significance of differences between means for percent muscle indicate that in every experiment the values for cultures maintained under the two conditions were significantly different ($p < 0.001$). Accompanying data from the same sequence of experiments are presented in Table 2. (From Hauschka and Konigsberg 1966)

sequent course of muscle differentiation. For example, what is the developmental fate of myoblasts not provided with a collagen substratum?

This question has been approached indirectly by comparing the plating efficiency and percent muscle differentiation between clonal populations grown in unconditioned medium, with and without a collagen substratum (HAUSCHKA and KONIGSBERG 1966). The data for such a series of experiments are presented in Tables 2 and 3. It is immediately apparent that although the same total number of colonies grew with or without the collagen substratum, the differentiation of individual colonies was remarkably affected by the two conditions. Looking specifically at the data from experiment 1, it can be seen that about 30 of the approximately 45 colonies per petri plate were scored as muscle when grown on collagen, whereas only 1 colony in 45 was scored as muscle in the absence of collagen. The remaining 29 "potential" muscle colonies (as predicted from the collagen substratum data) were apparently scored as fibroblast-like when collagen was not available to them. Thus roughly two thirds of the fibroblast-like colonies counted in the absence of collagen may really be muscle colonies in disguise. The other six experiments in this series may be similarly interpreted.

While the foregoing interpretation is the simplest which comes to mind, more complex alternatives should be mentioned. This explanation requires the dual assumption that collagen enhances the attachment, growth, and differentiation of muscle cells to precisely the same extent that it suppresses these functions in fibroblastic cells. Although sub-cloning experiments with fibroblasts indicate that such differential behavior is not characteristic of homogeneous cell populations, no unambiguous experiment to eliminate these possible interpretations for a heterogeneous primary cell suspension has yet been devised*.

If, as it seems likely, the initial interpretation is correct, what is the basis of the morphological alteration of the "potential" muscle colonies when grown in the absence of collagen? Specifically, one would like to know whether all of the fibroblast-like colonies appearing on untreated petri plates remain fibroblastic when subcloned on a collagen substratum, or whether some of them revert to myogenic behavior. A less ambiguous approach to this question can be achieved by performing an experiment similar to that diagrammed in Fig. 4.

Preliminary experiments of this type provide some information at the level of secondary clones, but are currently uninterpretable at the level of tertiary clones. The major shortcoming of these experiments is that colony morphology has served as the sole criterion for differentiation. Morphology is a useful, and probably adequate criterion for identification of colonies containing multinucleated muscle fibers; but it is inadequate for describing the extent of differentiation in colonies containing no obvious muscle fibers. The application of isotopic techniques to determine whether muscle-specific proteins or collagen are being synthesized by colonies grown under

---

* *Note Added in Proof.* Recent experiments designed to investigate the molecular mechanism of the collagen-myoblast interaction, indicate that collagen may exert an initial effect on cell binding *per se*. Thus when one compares the attachment of a suspension of embryonic leg muscle cells (containing a mixture of myoblasts and fibroblasts) to collagen-coated and uncoated petri plates, it is possible to distinguish a population of cells which fails to attach to the gelatin surface when transferred. We are presently attempting to identify the cell type which exhibits collagen-dependent binding.

4*

the different environmental conditions is critical to further interpretation. Nevertheless, the differential behavior of secondary clones in an experiment of the type outlined above, may have considerable relevance toward understanding the stability of myoblast differentiation.

Morphological results from such an experiment are shown in Fig. 5 and 6. Isolated muscle colonies were located with phase microscopy, trypsinized, and equal aliquots of cells were dispensed into collagen-coated and untreated petri plates. Two

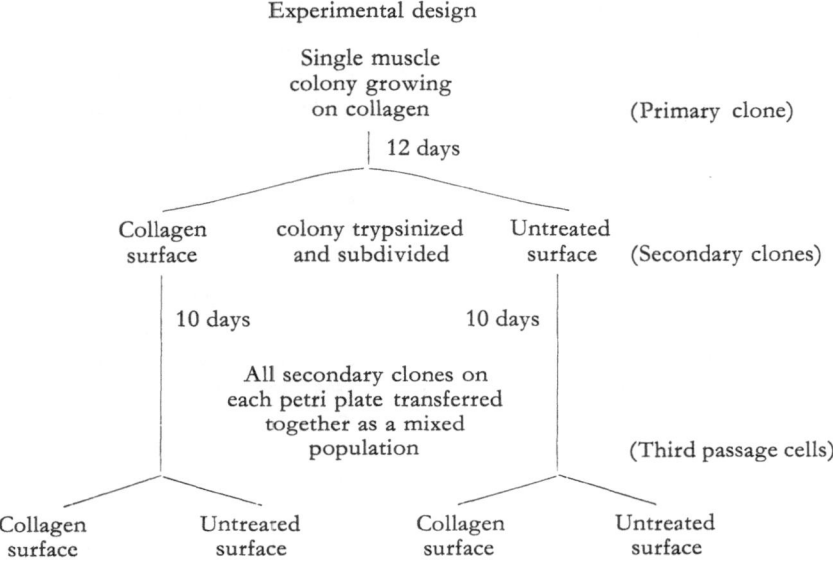

Fig. 4. Schematic design of an experiment for testing the effect of collagen or its absence on subcloned muscle cells. Results are discussed in the text. Photomicrographs of two representative subcloned populations are presented in Fig. 5 and 6. (From HAUSCHKA 1966)

weeks later, the secondary clones were fixed and stained. The appearance of such secondary colonies indicates that at least some of the second passage muscle cells maintain their ability to form multinucleated muscle fibers when provided with a collagen substratum; but this capacity remains unexpressed by genetically identical muscle cells cultured in the absence of collagen. Attempts to demonstrate retention of the capacity for muscle differentiation in third passage cells, among a population which has experienced even a 10 day sojourn in the absence of collagen, have proved fruitless to date. However, third passage muscle cells maintained on a collagen substratum throughout the primary and secondary phases of the experiment have likewise failed to develop multinucleated muscle fibers, and appear generally fibroblastic. Even secondary muscle clones exhibit decidely less tendency for extensive fiber formation; yet the elongated mononucleated cells of these colonies still orient themselves in the characteristic swirling pattern of muscle fibers.

The technical solution to these problems most likely lies in the length of time permitted to elapse between successive passage. Since those myoblasts which retain the capacity to form differentiated muscle, express this function by entering into

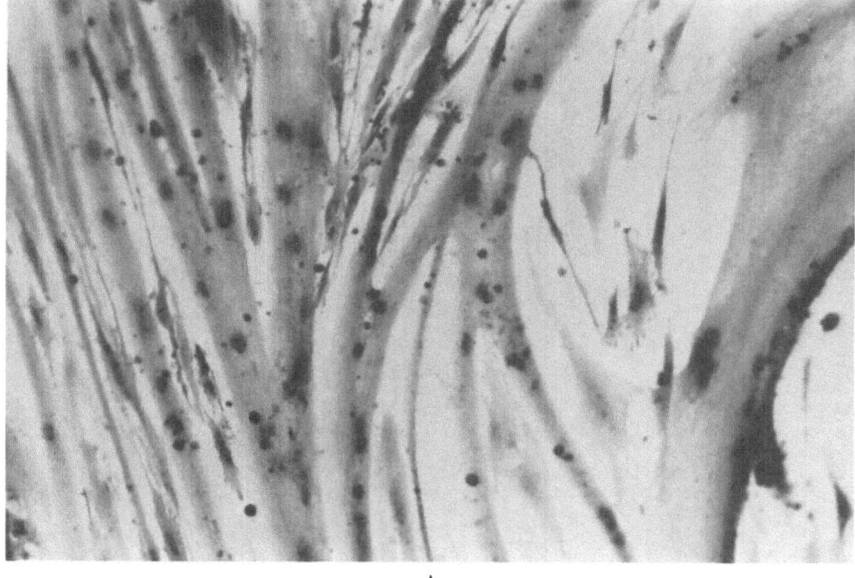

A

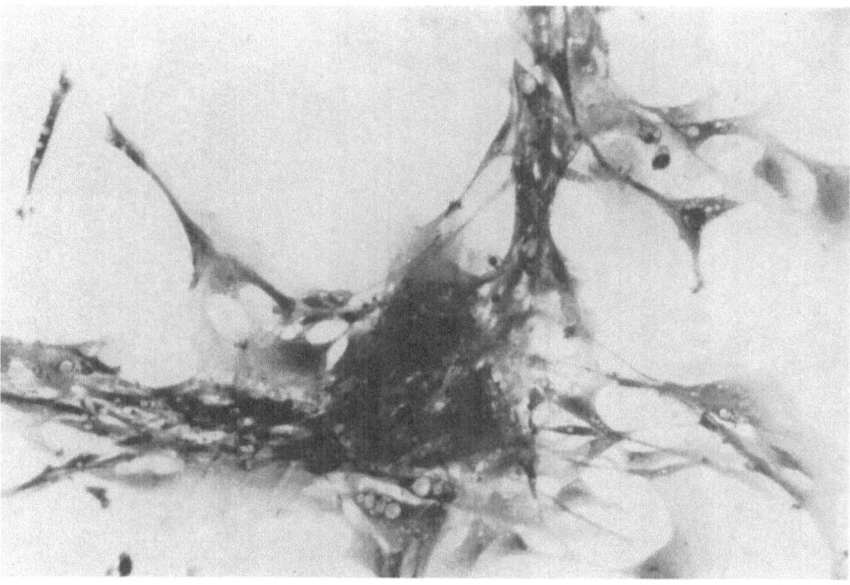

B

Fig. 5. Representative regions from subcloned muscle colonies grown in unconditioned medium. (A) Subclones grown on a collagen substratum; (B) Subclones grown on an untreated Petri plate surface. The subclones were initially derived from the same parental muscle colony as outlined in Fig. 4. The formation of muscle fibers was extensive in this particular subclone when the cells were exposed to collagen; but in the absence of collagen, slow-growing, extremely compact fibroblastic colonies appeared. Compare the behavior of this subclone to the subclone illustrated in Fig. 6

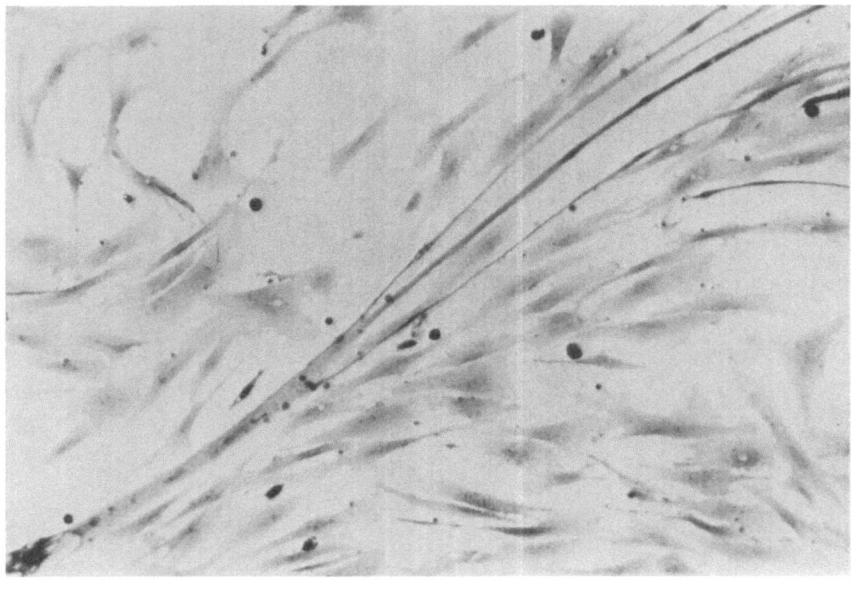

A

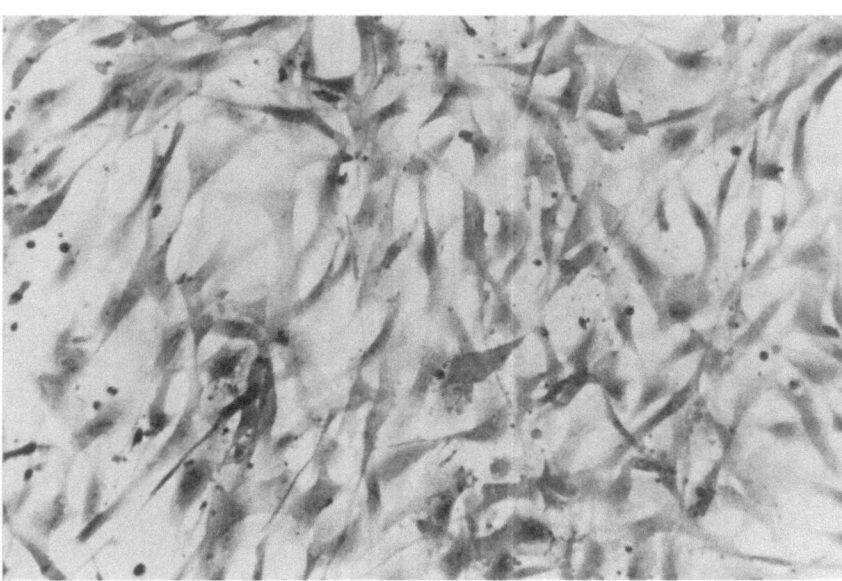

B

Fig. 6. Representative regions from another set of muscle subclones, treated identically to those shown in Fig. 5. This particular muscle colony gave rise to progeny colonies which exhibited little, if any, fusion when subcloned on a collagen substratum (A); and which grew as dispersed sheets of fibroblastic cells when subcloned in the absence of collagen (B). Compare the behavior of this subclone to the subclone illustrated in Fig. 5.

muscle fibers, thereby withdrawing themselves from the pool of proliferating cells, the design of the present experiment may unavoidably select against cells with the greatest fiber forming capacity. But how such diversity in cellular behavior is generated within a clonally derived population, is not clear at present.

The variability is made somewhat more confusing by the fact that although morphological differences exist between individual cells in most primary clones, the hundreds of secondary clones derived from a single muscle colony are remarkably uniform. Of course, they too exhibit morphological heterogeneity within themselves; but there is little, if any, tendency toward a "cloning out" of the various phenotypes discernible within the primary clone, suggesting that the differences observed are only transient oscillations of cellular morphology. A similar sort of reversible phenotypic behavior has been described following the exposure of muscle cultures to the thymidine analogue 5-bromodeoxyuridine; but in this instance, cell fusion was also inhibited (STOCKDALE et al. 1964; COLEMAN and COLEMAN 1966). Whether the fibroblastic cells present in normal muscle colonies have also lost their fusion capacity, has not yet been established. On the other hand, subclones derived from separate primary muscle clonies, often exhibit reproducibly different phenotypes (compare Fig. 5A to 6A and 5B to 6B), as if to suggest that genetic differences may already have existed between the progenitor myoblasts *in vivo*.

Clearly, such heritable differences could result either from alterations at the level of gene mutation, or at the level of gene activation. We have not yet been able to distinguish between these two possibilities; but, in principle, the problem is amenable to solution through the type of experiment outlined above. When coupled with biochemical markers such as the synthesis of tissue-specific proteins, this approach may very likely help elucidate the mechanism responsible for the transmission of selective gene activation from one cell generation to the next.

Acknowledgment

Much of the work reported in this paper was performed at the Carnegie Institution of Washington, Department of Embryology while I was a student research fellow. I am very grateful for the warm and generous support supplied by this group of people. I am particularly happy to acknowledge the stimulating companionship and collaboration of Dr. IRWIN R. KONIGSBERG and Mr. FRANCIS J. KUPRES, in whose laboratory the experiments were performed.

References

ALLEN, E. R., and F. A. PEPE: Ultrastructure of developing muscle cells in the chick embryo. Amer. J. Anat. **116**, 115—148 (1965).

ARESE, P., M. T. RINAUDO, and A. BOSIA: Levels of glycolytic intermediates in the musculature of the chick during embryonic and post-embryonic development. Europ. J. Biochem. **1**, 207—215 (1967).

BARIL, E. F., and H. HERRMANN: Studies of muscle development. II. Immunological and enzymatic properties and accumulation of chromatographically homogeneous myosin of the leg musculature of the developing chick. Develop. Biol. **15**, 318—333 (1967).

CAHN, R. D., and R. LASHER: Simultaneous synthesis of DNA and specialized cellular products by differentiating cartilage cells *in vitro*. Proc. nat. Acad. Sci. (Wash.) **58**, 1131—1138 (1967).

CAPERS, C. R.: Multinucleation of skeletal muscle in vitro. J. biophys. biochem. Cytol. **7**, 559—566 (1960).

COLEMAN, J. R., and A. W. COLEMAN: Reversible inhibition of clonal myogenesis by 5-Bromodeoxyuridine. J. Cell Biol. **31**, 22A; abstract No. 41 (1966).

Coleman, J.R., A.W.Coleman, and H.Roy: Myosin-containing mononuclear cells in myogenic cell cultures from chicken embryo leg muscles. Amer. Zool. 6, Abstr. No. 234 (1966).

Coon, H. G.: Clonal stability and phenotypic expression of chick cartilage cells *in vitro*. Proc. nat. Acad. Sci. (Wash.) 55, 66—73 (1966).

Cooper, W. G., and I. R. Konigsberg: Succinic dehydrogenase activity of muscle cells grown *in vitro*. Exp. Cell Res. 23, 576—581 (1961).

— — Dynamics of myogenesis *in vitro*. Anat. Rec. 140, 195—206 (1961).

Cosmos, E.: Enzymatic activity of differentiating muscle fibers I. Development of phosphorylase in muscles of the domestic fowl. Develop. Biol. 13, 163—181 (1966).

Dessouky, D. A., and R. G. Hibbs: An electron microscope study of the development of the somatic muscle of the chick embryo. Amer. J. Anat. 116, 523—566 (1965).

Deuchar, E. M.: Adenosine triphospatase activity in early somite tissue of the chick embryo. J. Embryol. exp. Morph. 8, 251—258 (1960).

Emmart, E. W., D. R. Kominz, and J. Miguel: The localization and distribution of glyceraldehyde-3-phosphate dehydrogenase in myoblasts and developing muscle fibers growing in culture. Histochem. Cytochem. 11, 207—217 (1963).

Eppenberger, H. M., R. von Fellenberg, R. Richterich, and H. Aebi: Die Ontogenese von zytoplasmatischen Enzymen beim Hühnerembryo. Enzymol. biol. cl·n 2, 139—174 (1962/63).

Firket, H.: Recherches sur la synthése des acides désoxyribonucléiques et la préparation à la mitose dans des cellules cultivées *in vitro* (Etude cytophotométric et autoradiographique). Arch. Biol. 69, 1—166 (1958).

— Ultrastructural aspects of myofibril formation in cultured skeletal muscle. Z. Zellforsch. 78, 313 (1967).

Fischman, D. A.: An electron microscope study of myofibril formation in embryonic chick skeletal muscle. J. Cell Biol. 32, 557—575 (1967).

Goodwin, B. C., and I. W. Sizer: Effects of spinal cord and substrate on acetylcholinesterase in chick embryonic skeletal muscle. Develop. Biol. 11, 136—153 (1965).

Green, H., and B. Goldberg: Collagen and cell protein synthesis by an established mammalian fibroblast line. Nature (Lond.) 204, 347—349 (1964).

Hanson, J., and J. Lowy: The structure of F-actin and of actin filaments isolated from muscle. J. molec. Biol. 6, 46—60 (1963).

Hauschka, S. D.: Previously unpublished data (1965—1966).

— Collagen and the differentiation of skeletal muscle *in vitro*. Ph. D. Thesis, Johns Hopkins University, Baltimore, Md. 1966.

—, and I. R. Konigsberg: The influence of collagen on the development of muscle colonies. Proc. nat. Acad. Sci. (Wash.) 55, 119—126 (1966).

— — and F. J. Kupres: Previously unpublished observations (1966).

Hay, E. D.: The fine structure of differentiating muscle in the salamander tail. Z. Zellforsch. Abt. Histochem. 59, 6—34 (1963).

Herrmann, H.: Studies of muscle development. Ann. N. Y. Acad. Sci. 55, 99—108 (1952).

—, and S. H. Barry: Accumulation of collagen in skeletal muscle, heart and liver of the chick embryo. Arch. Biochem. 55, 526—533 (1955).

Heywood, S. M., R. M. Dowben, and A. Rich: The identification of polyribosomes synthesizing myosin. Proc. nat. Acad. Sci. (Wash.) 57, 1002—1009 (1967).

Holtzer, H., J. M. Marshall, and H. Finck: An analysis of myogenesis by the use of fluorescent antimyosin. J. biophys. biochem. Cytol. 3, 705—724 (1957).

Huxley, H. E.: The double array of filaments in cross striated muscle. J. biophys. biochem. Cytol. 3, 631—648 1(957).

— Electron microscopic studies on the structure of natural and synthetic protein filaments from striated muscle. J. molec. Biol. 7, 281—308 (1963).

Kaplan, N. O., and R. D. Cahn: Lactic dehydrogenases and muscular dystrophy in the chicken. Proc. nat. Acad. Sci. (Wash.) 48, 2123—2130 (1962).

Kitiyakara, A.: The development of non-myotomic muscle of the chick embryo. Anat. Rec. 133, 35—45 (1959).

Konigsberg, I. R.: The differentiation of cross-striated myofibrils in short term cell culture. Exp. Cell Res. 21, 414—420 (1960).

KONIGSBERG, I. R.: Clonal analysis of myogenesis. Science 140, 1273—1284 (1963).

—, and S. D. HAUSCHKA: Cell and tissue interactions in the reproduction of cell type. In: *Reproduction: Molecular, subcellular, and cellular* (M. LOCKE, ed.), pp. 243—290. New York: Academic Press 1965.

—, and H. HERRMANN: The accumulation of alkaline phosphatase in developing chick muscle. Arch. biochem. 55, 534—545 (1955).

—, and F. J. KUPRES: Previously unpublished observations (1965—1966).

—, N. McELVAIN, M. TOOTLE, and H. HERRMANN: The dissociability of deoxyribonucleic acid synthesis from the development of multinuclearity of muscle cells in culture. J. biophys. biochem. Cytol. 8, 333—343 (1960).

LASH, J. W., H. HOLTZER, and H. SWIFT: Regeneration of mature skeletal muscle. Anat. Rec. 128, 679—693 (1957).

LEWIS, W. H., and M. LEWIS: Behavior of cross-striated muscle in tissue culture. Amer. J. Anat. 22, 169—194 (1917).

LOCKWOOD, D. H., F. E. STOCKDALE, and Y. J. TOPPER: Hormonedependent differentiation of mammary gland: Sequence of action of hormones in relation to cell cycle. Science 156, 945—946 (1967).

MARCHOK, A. C., and H. HERRMANN: Studies of muscle development I. Changes in cell proliferation. Develop. Biol. 15, 129—155 (1967).

MINTZ, B., and W. W. BAKER: Normal mammalian muscle differentiation and gene control of isocitrate dehydrogenase synthesis. Proc. nat. Acad. Sci. (Wash.) 58, 592—598 (1967).

MOOG, F.: Adenylpyrophosphatase in brain, liver, heart, and muscle of chick embryos and hatched chicks. J. exp. Zool. 105, 209—220 (1947).

NAMEROFF, M., and H. HOLTZER: The loss of phenotypic traits by differentiated cells. IV. Changes in polysaccharides produced by dividing chondrocytes. Develop. Biol. 16, 250—281 (1967).

OBINATA, T., M. YAMAMOTO, and K. MARUYAMA: The identification of randomly formed thin filaments in differentiating muscle cells of the chick embryo. Develop. Biol. 14, 192—213 (1966).

OKAZAKI, K., and H. HOLTZER: Myogenesis: Fusion, myosin synthesis, and the mitotic cycle. Proc. nat. Acad. Sci. (Wash.) 56, 1484—1490 (1966).

PETTE, D., W. LUH, and TH. BÜCHER: A constant-proportion group in the enzyme activity pattern of the Embden-Meyerhof chain. Biochem. biophys. Res. Commun. 7, 419—424 (1962).

—, M. KLINGENBERG, and TH. BÜCHER: Comparable and specific proportions in the mitochondrial enzyme activity pattern. Biochem. biophys. Res. Commun. 7, 425—429 (1962).

PRZYBYLSKI, R. J., and J. M. BLUMBERG: Ultrastructural aspects of myogenesis in the chick. Lab. Invest. 15, 836—863 (1966).

REPORTER, M. C., I. R. KONIGSBERG, and B. L. STREHLER: Kinetics of accumulation of creatine phosphokinase activity in developing embryonic skeletal muscle *in vivo* and in monolayer culture. Exp. Cell Res. 30, 410—417 (1963).

SHAFIG, S. A.: Electron microscopic studies on the indirect flight muscles of Drosophila melanogaster. II. Differentiating myofibrils. J. Cell Biol. 17, 363—373 (1963).

STOCKDALE, F. E., and H. HOLTZER: DNA synthesis and myogenesis. Exp. Cell. Res 24, 508—520 (1961).

—, J. ABBOTT, S. HOLTZER, and H. HOLTZER: The loss of phenotypic traits by differentiated cells. II. Behavior of chondrocytes and their progeny *in vitro*. Develop. Biol. 7, 293—302 (1963).

—, K. OKAZAKI, M. NAMEROFF, and H. HOLTZER: 5-Bromodeoxyuridine: Effect on myogenesis *in vitro*. Science 146, 533—535 (1964).

—, and Y. J. TOPPER: The role of DNA synthesis and mitosis in hormone-dependent differentiation. Proc. nat. Acad. Sci. (Wash.) 56, 1283—1289 (1966).

YAFFE, D., and M. FELDMAN: The formation of hybrid multinucleated muscle fibers from myoblasts of different genetic origin. Develop. Biol. 11, 300—317 (1965).

—, and S. FUCHS: Autoradiographic study of the incorporation of uridine — ^{3}H during myogenesis in tissue culture. Develop. Biol. 15, 33—50 (1967).

Factors Affecting Inheritance and Expression of Differentiation: Some Methods of Analysis

ROBERT D. CAHN

Department of Zoology, University of Washington, Seattle, Washington 98105

I. Introduction

Differentiated cells grow, develop and maintain their specialized properties during normal embryogenesis (MANASEK 1968a, b). In certain tissues (e.g. skeletal muscle, pancreas, mammary gland, and hematopoietic tissues) the cells do not synthesize large amounts of specialized products until DNA replication and rapid cell division have ceased. Nonetheless, even in these tissues one can detect small but measurable amounts of their specialized proteins during the rapid growth phase of the stem cells (RUTTER et al. 1967; WESSELLS 1967).

Other cells, such as cartilage and pigmented retina (P R) appear to synthesize considerable, but perhaps not maximal, amounts of their specialized products even during rapid growth (CAHN and LASHER 1967). It has been found, however, that normally *in vivo* the amount of differentiated product present in the cells is small until cell growth has slowed or the cells have ceased dividing altogether (COLOUMBRE 1955).

Data from experiments performed on cells differentiating *in vivo* which would tell us whether or not cells can replicate DNA and divide rapidly while maintaining appreciable synthesis of their specialized products are scanty. Attempts to study this problem during cell and organ growth and differentiation *in vitro* have usually led to the conclusion that rapidly dividing cells, cells which are rapidly replicating their DNA, do not synthesize appreciable amounts of specialized cellular products (HOLTZER et al. 1960). Thus the embryological dictum: "Dividing cells don't differentiate".

However, in recent years evidence has been accumulating that under certain conditions some cells can divide rapidly while maintaining the ability to synthesize appreciable amounts of specialized cellular products. These reports (COON 1966; CAHN and CAHN 1966; CAHN and LASHER 1967; COON and MARZULLO 1967; MANASEK 1968) have demonstrated that cartilage and pigmented retina cells can, in the proper environment, divide rapidly while maintaining their morphologically differentiated characteristics. COON (1964) and COON and MARZULLO (1967) have identified the sulphated polysaccharide substances synthesized by the differentiated cartilage cell clones as chondroitin sulphates A and C.

These and more recent studies (COON and CAHN 1966) have demonstrated that the cellular micro-environment plays a crucial role in the outcome of these experiments. PR cells placed in certain culture media do not express melanin and melanosome synthesis. Cartilage cells also do not express their differentiation in many media. Nonetheless, cells cultured as undifferentiated clones or monolayers for many serial passages can express their differentiation when replaced in appropriate culture conditions.

Questions. These results raise several questions concerning the mechanisms for stabilization of differentiation:

(1) What environmental factors are needed for maximal *expression* of a cell's innate or "cryptic" (GROBSTEIN 1959, 1964) differentation?

(2) Does rapid cell division and/or DNA replication exclude simultaneous synthesis of differentiated products?

(3) What molecular mechanisms does a cell employ to *store* information as to its state of differentiation, and how is this released during overt expression of the differentiation?

This discussion will address itself to these three problems in turn.

Terminology. The bias of a cell towards a particular type of differentiation is heritable during at least many tens of cell divisions (COON 1966; CAHN and CAHN 1966). This *heritable ability* to differentiate into a particular cell type will be called the cell's *epigenotype.* These root terms, familiar to embryologists, have been used in several other words both as an adjective by WADDINGTON ("Epigenetic") and in a slightly different permutation ("Epigenesis") by others to connote the entirety of regular and sequential developmental events. *Epigenotype* here connotes a stable heritable character, whose mode of impression, as yet not known, is on top of or in addition to the classically understood genotype, i.e. the DNA base sequences.

A cell may *inherit* the ability to differentiate during long periods, yet remain cryptic or unexpressed with regard to its characteristic cytodifferentiation, i.e. its specific protein and non-protein products and structures. Such "unexpressed" cells may possess a few molecules of these characteristic products, but lack the proper environmental cues for expression of maximal and recognizable cytodifferentiation. I will refer to cells in such a state as *unexpressed.*

The aggregate of synthetic capabilities and other properties resulting from the expression of recognizable cytodifferentiation will be referred to as a cell's *epiphenotype.*

II. Role of Environment in Regulating Expression of Epiphenotype

Chick cells freshly liberated from tissues of the intact organism and grown as confluent monolayers in the usual media containing varying proportions of serum, chick embryo extract and nutrient salt mixture do not usually express their epiphenotype. I would like to consider briefly the effects of several of these environmental variables on the expression of epiphenotype, deferring until later a consideration of what the underlying mechanisms of such effects may be.

A. Serum

Sera vary greatly in their ability to promote differentation of various cell types. Horse serum, almost a *sine qua non* for muscle differentiation (multinucleated myotubes) is in most cases either toxic or allows only poor differentiation of cartilage and pigmented retina cells. Various fetal calf sera (FCS) also differ significantly in their ability to allow expression of the cartilage or pigmented retina epiphenotype. This is illustrated in Table 1a. One serum may be good for one tissue and poor for another, or *vice versa.* As will be shown in the next section, this difference amongst sera does not appear to be due to any difference in their ability to support growth of the different tissues. Growth rates in various fetal calf sera differing markedly in their ability to promote differentiation are usually quite comparable. These effects of various fetal calf sera may be due to bound micronutrients, or, more probably, to varying hormone or specialized protein levels, as will be discussed below.

Table 1a. *Differential effects of various lots of fetal calf serum on plating efficiency and differentiation of chick embryo cartilage and pigmented retina cells*

Serum lot number	Cartilage cells % Plating efficiency	% Differentiated clones	Pigmented retina cells % Plating efficiency	% Differentiated clones
62309 C	11	4.5	25.7	0.3
62309 E	19	37	31.0	18
62309 F	22	18	34.3	48.3
62309 D	23	17	32.7	52.6
62209 M	27	41	31.0	45.2
62209 K	53	70	32.0	47.0
62209 L	42	64	30.1	59.3
62209 H	39	59	24.0	49.8
62209 J	55	55	28.8	43.8
62209 G	68	68	26.8	41.6

All cells were plated in medium containing 5% of the respective fetal calf serum and 1% of the H fraction of embryo extract. 4 plates were counted for each point. The standard deviations were approximately 10—12% of the mean values reported here. Each point represents approximately 100—600 colonies counted. All sera were made into medium on the same day, and testing was done simultaneously. Sera were from Grand Island Biological Co.

Table 1b. *Comparison of effects of chick and quail embryo extracts on growth, plating efficiency and expression of differentiation of chick and quail embryo chondrocytes*

Medium	Clonal cultures[a] % CMC Chick cells	Quail cells	% P.E. Chick cells	Quail cells	Mass monolayer cultures[b] Chick cultures Growth[e]	Differentiation[d]	Quail cultures Growth	Differentiation
1% Quail EE	83.7	49.0	28.1	3.9	2.2, 2.8	none	3.1, 2.9	+
1% Chick EE	78.9	30.1	29.2	2.8	3.1, 3.2	none	3.1, 3.3	+ +
5% Quail EE	0	[e]	0.02	[e]	3.4, 3.5	none	4.2, 4.8	+
5% Chick EE	28.8	[e]	0.4	[e]	3.1, 2.9	none	4.0, 4.4	+ +

CMC = cartilage making colonies.
P.E. = plating efficiency.
EE = Embryo extract (Prepared as in Cahn 1967).
[a] 1 000 cells were plated on a 100 mm preti dish.
[b] 10[6] cells were plated on a 100 mm petri dish.
[e] after 4 days cells were trypsinized from dish; numbers represent cells per dish $\times$ 10^{-6}.
[d] Plates were fixed, stained with toluidine blue and scanned for centers of metachromasia.
    + = 4—6 "nodules" with matrix material per microscope field.
+ + = 9—14 "nodules" with matrix material per microscope field.
[e] The quail embryo extract was toxic at 5%. No clones survived.

B. Basic Nutrient Media

Cartilage and pigmented retina differ markedly in their ability to express in various basic nutrient media. This is illustrated in Table 2. Even the fairly minor differences in medium composition such as exist between F_{10} and F_{12} produce significant differences in cell epiphenotype. Media prepared in the author's laboratory are invariably superior to those purchased from commercial sources.

C. Embryo Extract

As we have reported previously (COON and CAHN 1966) embryo extract (EE) also has profound effects on expression of cartilage and pigmented retina cells. Recently, using 0.1–0.5 percent of the Sephadex G-25 or G-150 excluded fraction of chick EE, we have been able to obtain optimal growth and differentiation of both these two cell types. The requirement for EE may be eliminated entirely with certain fetal calf sera used at 10–15% (Table 2). It is interesting to note, however, that quail cells respond quite differently to the same concentrations of chick EE which cause dedifferentiation of chick cells (Table 1 b). Quail chondrocytes differentiate even in 10 percent EE.

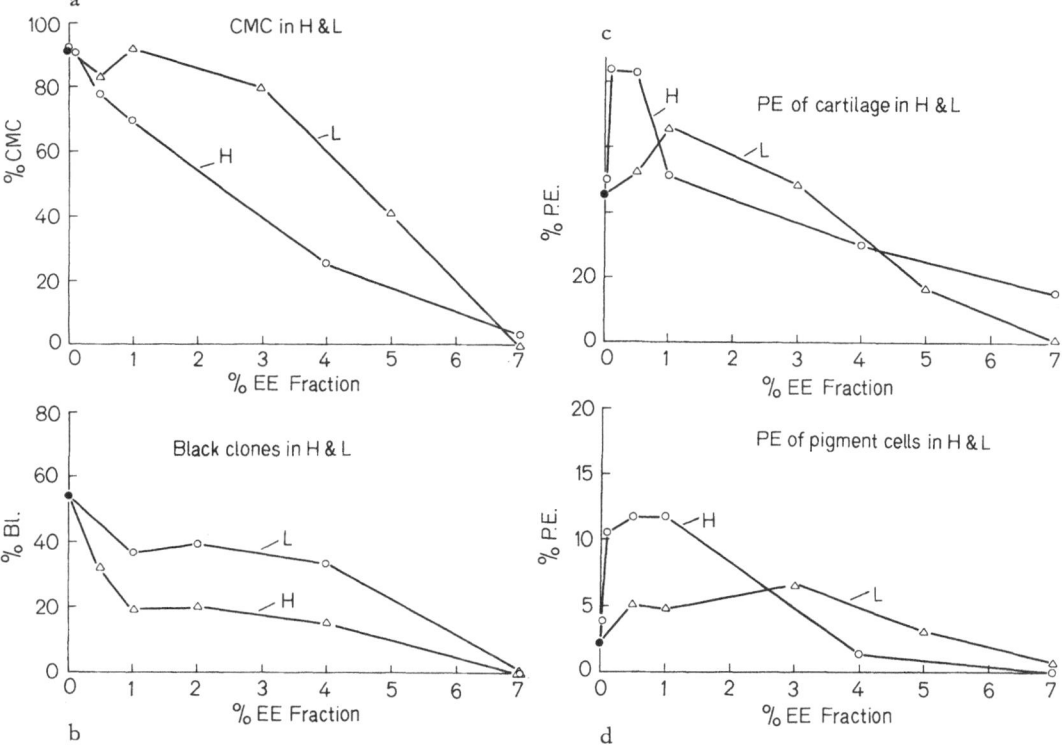

Fig. 1. Effects of chick embryo extract Sephadex G-25 fractions (H = heavy; L = light) on expression of differentiation and plating efficiency of chick cartilage and pigmented retina cells. a) Change in % cartilage making colonies (CMC) with increasing % EE fractions. b) Change in % colonies appearing black to naked eye with increasing % EE. c) Change in plating efficiency of cartilage with increasing % EE fractions. d) Change in plating efficiency of pigmented retina with increasing % EE

There does not appear to be any profound qualitative species specificity to the effect, as quail EE has effects similar to chick EE on both cell types. Cartilage and pigmented retina differ also in their quantitative response to embryo extract (Fig. 1).

Table 2. *A comparison of the effects of Ham's media F_{10} and F_{12} on chick embryo cartilage plating efficiency and expression of differentiation*

Medium additions	Plating efficiency		% Cartilage making clones	
	F_{10}	F_{12}	F_{10}	F_{12}
5% FCS 0.1% H-fraction of EE	92.5 ± 16.7	101.8 ± 5.8	47.5 ± 2.9	63.4 ± 4.5
5% FCS only	18.0 ± 6.6	26.0 ± 4.9	50.3 ± 13.1	55.8 ± 11.9
10% FCS only	63.4 ± 10.6	65.2 ± 7.6	84.2 ± 6.9	86.5 ± 2.4
15% FCS only	81.6 ± 6.0	82.0 ± 12.9	89.6 ± 4.4	90.7 ± 4.1

Although the differences are small in some cases, statistically highly significant in others, cells grown in medium F_{12} *always* have a higher plating efficiency and a higher % expression of differentiation under the conditions of this experiment. It is also interesting to note that fetal calf serum (FCS) can substitute for the plating efficiency promoting factor (s) in embryo extract (EE), and can promote expression of differentiation in a manner superior to EE containing media

D. Hormones

Hormonal effects on cells in culture have been reported in several cell systems (Cox and MACLEOD 1962; FELL 1967). Recently COLEMAN (1967) has reported that thyroxin treatment allows mass monolayers of chick cartilage to differentiate. In confirming and extending this observation, we have found a significant effect of thyroxin in promoting differentiation of pigmented retina cell clones. A series of experiments is in progress testing other purified hormones for differentiation promoting activity.

It should be noted that many different hormones may be present in chick EE. Insulin is present in 9 day chick EE (CAHN and CLARK, unpublished). The immunologically identified hormone (BERSON and YALOW 1959a, b) fractionates with the Sephadex G-150 excluded fraction of the EE. In experiments testing hormone effects assays should be carried out so as to detect the presence of other hormones in a bound condition.

FELL (1967) has recently reported striking effects of hydrocortisone and ε-aminohexanoic acid on differentiation of cartilage *in vitro*. Both these compounds can reverse the "dedifferentiation" of cartilage caused by organ culture of bone rudiments in marginally deleterious but non-toxic concentrations of certain types of antisera. It is FELL's suggestion that many of the effects of these two compounds can be explained on the basis of their action to prevent lysosomal breakdown. A membrane locus was also suggested as the site of action of hydrocortisone by MELNYKOVYTCH (1966). Certainly the plasma membranes of cells in clonal and monolayer culture are exposed to the rigours of the external environment to a degree not found in the intact tissue. For this reason "non-physiological" amounts of some hormones or other surface active compounds may be required to "toughen" the cell surface and prevent activation of the sublethal cytolytic effects resulting from lysosomal breakdown. Intracellular release of lysosomal hydrolases under such conditions may well account for the lack of differentiation under most conditions by the most peripheral cells in clones of pigmented retina and cartilage (see pictures in CAHN and CAHN 1966; CAHN, COON, and CAHN 1968) in some sera and in certain mass cultures, rather than the frequently offered explanation that these cells are not differentiating because they are dividing too rapidly.

E. Effects of Drugs

In the course of studies carried out over the last two years with the intention of obtaining drug resistant mutants of differentiated chick cells we noticed that very low concentrations of various drugs ($10^{-7} M$), concentrations which did not kill the cells, nonetheless prevented cartilage and pigmented retina cells from differentiating (LASHER and CAHN, unpublished). 5-Bromodeoxyuridine (BUDR), 8-Azaguanine, and Nitrosoguanidine all were effective, although they had interesting differences in their effects on the morphology of the resultant cells. Similar effects of BUDR in myoblast fusion have been previously reported by STOCKDALE et al. (1964) and by COLEMAN and COLEMAN (1966).

COLEMAN reported that in monolayer cultures and clones of embryonic myoblasts, fusion to form multinucleated myoblasts is inhibited by BUDR, but the individual myoblasts recover again after the drug is removed. It is not clear from his studies whether the cells need to grow after removal from the drug, as they need only fuse to make myotubes. However, he has presented evidence (COLEMAN 1967) that, released from BUDR, they do synthesize DNA. In similar clonal cultures of chondrocytes, we have found several different types of effects of BUDR. The results are summarized in Tables 3, 4 and 5.

1. 10^{-5}, 10^{-6}, and $10^{-7} M$ BUDR added to ordinary Ham's F_{10} medium with serum supplements but lacking thymidine, completely prevents expression of the differentiation of cartilage cell clones. When the initial plating densities are greater than 10^3 cells on a 60 mm dish (5 ml medium) the loss of differentiation appears to be at least partially reversible after even as long as 9 days exposure to BUDR. When less than 100 cells are inoculated, this long exposure to BUDR causes irreversible loss of differentiation and eventually of viability, except in the lowest concentrations of BUDR ($10^{-7} M$). This should be kept in mind when interpreting the results of ABBOTT and

Table 3. *Effects of 5-Bromo-deoxyuridine (BUDR) on cartilage cell differentiation: Continuous growth in BUDR*

Passage Number	Plating density	Number of plates	Medium	BUDR concentration	% CMC*	% N-CMC*	%FB*	% P.E.*
I	200	4	+ thy	—	86.2 ± 1.8	6.8 ± 1.6	7.1 ± 0.9	91.4 ± 9.1
I	200	4	— thy	$10^{-8} M$	81.3 ± 3.4	5.2 ± 2.4	13.5 ± 1.2	82.0 ± 9.7
II	200	3	— thy	—	70.5 ± 5.1	18.9 ± 3.6	10.6 ± 2.4	46.0 ± 6.1
II	200	1	+ thy	—	88.0	9.6	2.4	62.5
II	200	4	— thy	$10^{-5} M$	0.0	0.0	100.0	39.1 ± 2.9
II	200	4	— thy	$10^{-6} M$	0.0	0.0	100.0	41.5 ± 4.9
II	200	4	— thy	$10^{-7} M$	0.0	2.0 ± 1.2	98.0 ± 1.2	44.3 ± 1.8

CMC = Cartilage making clones.
N-CMC = Epithelial shaped non-cartilage making colonies.
FB = Fibroblastic appearing colonies.
P.E. = plating efficiency.
* = $\pm$ standard deviation amongst the plates counted.

Table 4. *Effect of low concentration of 5-Bromo-deoxyuridine (BUDR, $10^{-7}M$) on cartilage cell differentiation: partial growth in BUDR*

Passage number	Plating density	Number of plates	Medium	Time in BUDR	Medium after BUDR	% CMC*	% N-CMC*	% FB*	% P.E.*
I	200	4	+ thy	—	—	86.2 ± 1.8	6.8 ± 1.6	7.1 ± 0.9	91.4 ± 9.1
I	200	4	— thy	1 day	+ thy	80.5 ± 4.3	7.4 ± 1.9	13.6 ± 0.3	83.5 ± 5.0
I	200	4	— thy	2 days	+ thy	74.0 ± 2.2	14.3 ± 3.1	13.2 ± 1.3	87.0 ± 4.9
I	200	4	— thy	3 days	+ thy	48.1 ± 7.3	46.0 ± 12.6	8.2 ± 2.7	83.8 ± 4.3
II	200	4	— thy	—	—	42.2 ± 5.3	50.6 ± 3.6	7.0 ± 3.2	24.1 ± 5.6
II	200	4	— thy	4 days	— thy	31.7 ± 6.1	7.1 ± 3.7	61.2 ± 7.4	29.1 ± 4.8
II	200	4	— thy	5 days	— thy	8.6 ± 3.9	2.3 ± 1.4	89.0 ± 4.7	27.1 ± 4.3
II	200	4	— thy	6 days	— thy	0.4 ± 0.7	1.6 ± 2.0	98.0 ± 1.8	27.2 ± 3.9

CMC = Cartilage making clones.
N-CMC = Epithelial shaped non cartilage making colonies.
FB = Fibroblastic appearing colonies.
* = ± standard deviation amongst the 4 plates counted.
P.E. = plating efficiency.

Table 5. *Reversal of the effect of 5-Bromo-deoxyuridine (BUDR) in presence of uridine and thymidine*

Plating density	Medium additions BUDR	Thym.	Urid.	% CMC*	% N-CMC*	% FB*	% P.E.*
1000ᵃ	—	—	—	59.2 ± 8.5	38.6 ± 9.2	2.2 ± 0.9	20.7 ± 4.5
1000ᵃ	—	+	—	72.1 ± 1.4	25.9 ± 1.3	1.9 ± 0.8	26.6 ± 1.5
1000ᵃ	—	—	+	69.3 ± 1.6	30.6 ± 1.7	0.1 ± 0.2	26.0 ± 2.3
1000ᵃ	+	—	—	0.6 ± 0.2	5.7 ± 2.3	93.7 ± 2.5	20.1 ± 0.9
1000ᵃ	—	+	+	81.6 ± 1.9	17.5 ± 2.1	0.9 ± 0.2	27.3 ± 1.3
1000ᵃ	+	+	—	74.9 ± 1.2	21.2 ± 0.6	3.9 ± 0.9	30.4 ± 5.7
1000ᵃ	+	—	+	75.4 ± 2.3	22.6 ± 0.7	1.2 ± 0.5	31.2 ± 1.1
1000ᵃ	+	+	+	82.1 ± 2.4	17.6 ± 2.2	0.2 ± 0.3	29.0 ± 1.5
200ᵇ	—	—	—	42.4 ± 5.3	50.6 ± 3.6	7.0 ± 3.2	24.1 ± 5.6
200ᵇ	—	+	—	65.5 ± 8.0	31.2 ± 5.9	3.3 ± 3.1	30.0 ± 3.7
200ᵇ	—	—	+	50.2 ± 8.5	47.8 ± 7.5	2.2 ± 1.7	25.6 ± 4.0
200ᵇ	+	—	—	0.5 ± 0.0	32.3 ± 21.6	67.2 ± 21.4	18.6 ± 3.3
200ᵇ	—	+	+	80.2 ± 4.0	17.7 ± 4.8	2.2 ± 1.9	29.4 ± 2.9
200ᵇ	+	+	—	57.6 ± 6.6	35.3 ± 9.9	7.1 ± 3.2	26.0 ± 4.4
200ᵇ	+	—	+	55.5 ± 6.0	38.6 ± 6.4	5.9 ± 2.9	29.0 ± 4.9
200ᵇ	+	+	+	74.4 ± 3.0	24.4 ± 2.5	1.2 ± 0.7	32.3 ± 2.6

4 plates were counted in each experiment. ᵃ and ᵇ represent two different series. BUDR = $10^{-7}M$; uridine (Urid.) = $10^{-3}M$; thymidine (Thym.) = $3.0 \times 10^{-6}M$. Other abbreviations and symbols as in Tables 3 and 4. The cells were all II passage.

Fig. 2. Appearance of colonies of chick embryo chondrocytes with increasing duration of exposure to 5-Bromo-deoxyuridine (BUDR). a) Cartilage making colony. Left: piled up area in center of fixed and stained colony, hematoxylin and eosin, 100 × magnification. Right: live phase contrast view of small, well differentiated CMC near edge of clone. 250 × magnification. b) Epithelial type colony, short exposure to BUDR. Left: typical epithelial colony derived from cartilage making colony. Cells have not piled up in center. 100 × magnification. Right: Intermediate type colony, partially fibroblastic, partially epithelial. 100 × magnification. c) Fibroblastic type colony. Left: elongate aligned cells of typical colony. 100 × magnifiaction. Right: enlargement of center of fibroblastic colony, 200 × magnification

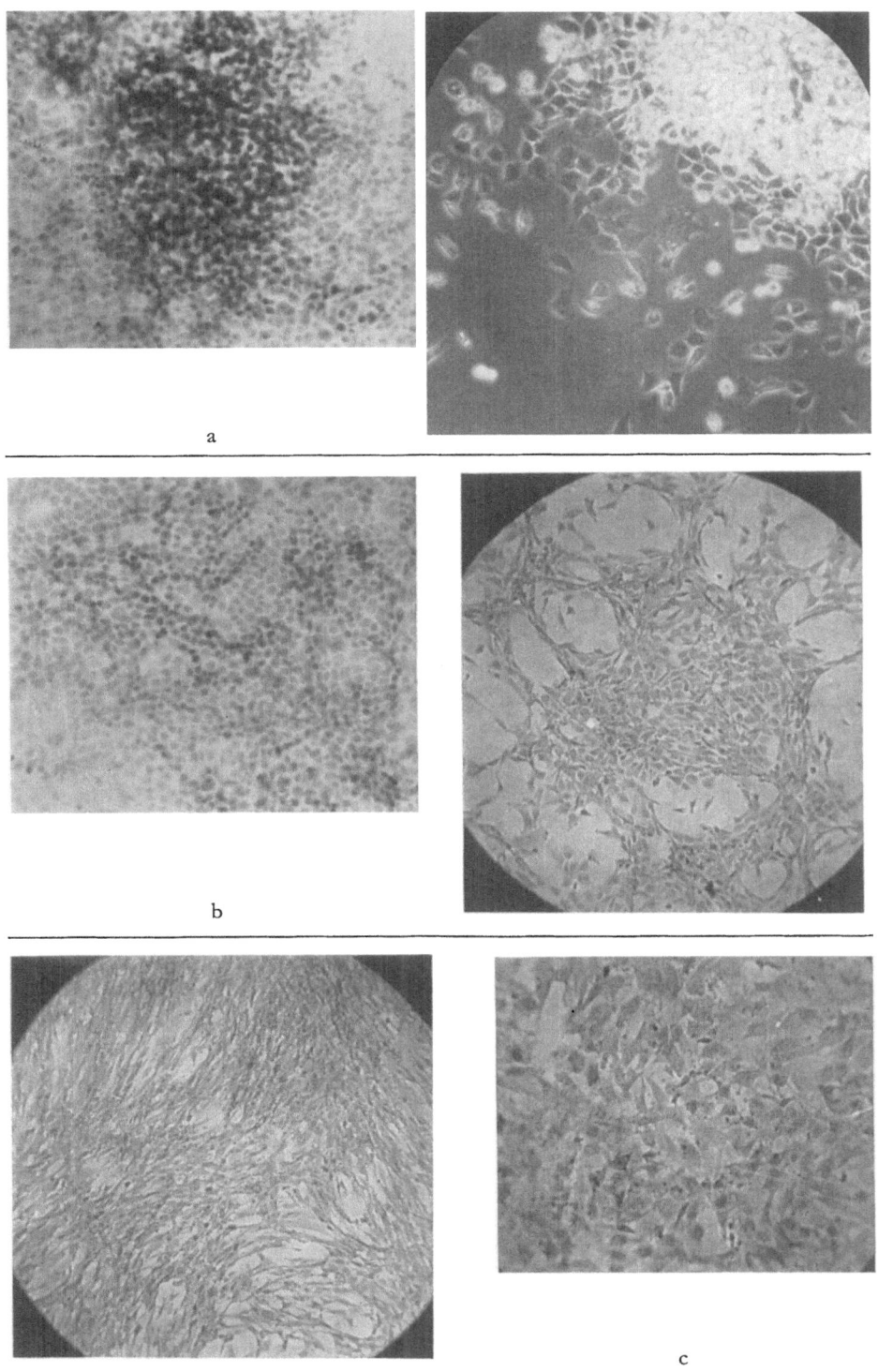

a

b

c

Holtzer (1968, and p. 1—16 in this volume) in which they reported the effects of BUDR on cartilage cells. In their experiments 50 μg/ml BUDR were used, which is greater than 10^{-4} M. Under similar experimental conditions in our laboratory, most of the cells become giant cells quite rapidly, and their differentiation is irreversibly inhibited even at high densities of initial cell platings.

2. Treatment of chondrocytes with 10^{-7} M BUDR for less than three days allows good recovery of colony forming ability and differentiation of the resultant clones. Treatment for longer times results in a marked drop in the number of differentiating colonies, but does not markedly affect the plating efficiency.

3. The effect of BUDR on differentiation is completely reversed by thymidine, uridine, and a combination of both nucleosides. Addition of uridine seems to allow the cells to differentiate better than uridine-free standard medium. On the other hand, lack of thymidine may reduce somewhat the number of differentiated colonies, but does not have anywhere near the magnitude of the effect of BUDR.

4. A detailed analysis of the data shows that the first effect of BUDR is to transform the cells into epithelial cells which, however, cannot make chondroitin sulfate (see Fig. 2b). The same morphological effect is caused by short (10 minute) exposures of the cells to nitrosoguanidine. Longer exposure to BUDR converts all the cells into a very "fibroblastic" cell type (see Fig. 2c), which appears to grow at least as well if not faster than the original cell type. Such cells are not truly "transformed" in the sense that they have become cancer cells or stable cell lines.

5. "Fibroblasts" picked from 4 or 5 day cultures grown in BUDR, when recloned in normal media or media with uridine give rise to colonies which show all three types of morphology and differentiation, from fully differentiated CMC to fibroblasts. Thus it appears that the shorter exposures to the drug cause a partially reversible change in ability to differentiate. Longer exposures cause changes in the cells which are as yet not completely reversible upon recloning.

At present it is not feasible to speculate extensively on possible molecular explanations for these results, but one could propose that differentiation relies on the production of an RNA or DNA species, whose production is preferentially interfered with by low concentrations of BUDR. Functions concerning cell maintenance and those functions necessary for rapid cell growth are not hit by these low concentrations of BUDR, at least not initially. Whether such entities are self-replicating, so that when their concentration is reduced to zero by cell multiplication in the absence of their replication the cell is permanently "dedifferentiated" cannot be adduced from the data at hand. This possibility is akin to the early suggestion by Spiegelman that differentiation is due to stable self-replicating "plasmagenes". The answer to this and other tentative hypotheses will have to await further experimentation in this field.

F. Effects of Cell Source, Shape and Density on Ability to Express

Earlier reports from several laboratories including our own (Cahn and Cahn 1966; Coon 1966; Whittaker 1963, 1967; Ephrussi and Temin 1960; Holtzer et al. 1960) indicated that under the usual culture conditions, high density mass monolayer cultures of cartilage and pigmented retina differentiate poorly. This is not always the case, however. EPHRUSSI and TEMIN state that the iris epithelium cultures eventually pigment after long periods at saturation densities. We have recently made two observations bearing on this point. 1) *Quail* cartilage cells differentiate even at initial plating densities of 10^6 cells per 60 mm petri plate, in media containing 5% or 10% EE. 2) In media containing 15% FCS *chicken* pigmented retina mass cultures (greater than 10^6 cells per plate) can differentiate well, although the amount of pigment per cell is considerably less than in clones. Under similar conditions, or as clones, *quail*

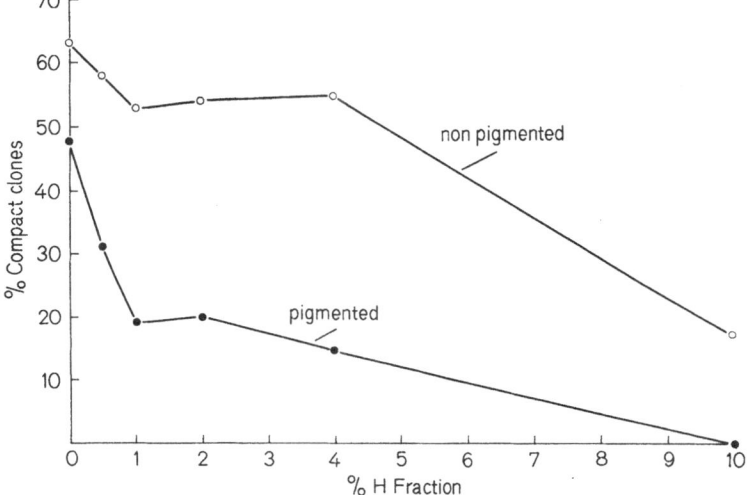

Fig. 3. Effect of H fraction of embryo extract on clonal morphology of pigmented retina cells. Cells retain their epithelial morphology at concentrations of embryo extract which cause complete loss of expression of pigmentation of the cells

pigmented retina cells differentiate poorly if at all. An additional observation to be borne in mind is that of J. R. COLEMAN (personal communication) which indicates that most cartilage cultures can differentiate much better under *lowered* O_2 tension and in the presence of thyroxine.

Media which promote the loss of differentiation by pigmented retina and cartilage cells also alter the shape of the cells. It is difficult to design experiments to test cause and effect relationships here, but some inferences can be drawn from the data shown in Fig. 3. It can be seen that the quantitative response of cell shape and overt pigment differentiation to EE is different. It is possible to have cultures of P R or cartilage which retain their "differentiated like" cuboidal shape, yet do not synthesize other recognizable cell specific products. The consistent observation has been made by several investigators however, that the cells which do not differentiate on the edge of colonies are usually excessively flattened, elongated and often "fibroblast-like". The generalization can thus be made that elongated flattened cells usually do not differentiate; but the reverse is not true: normal "differentiated-like" morphology is not a sufficient condition for maximal expression of cell specific product synthesis. Similar conclusions can be drawn from the sequence of events in "dedifferentiation" caused by BUDR. It is possible that the generally poor differentiation of highly flattened cells is due to the very high surface-volume ratio. Cells may lose specialized products to

5*

the medium faster than they can accumulate them. The excessive exposure to the environment may have other deleterious effects as well.

It is difficult to integrate this body of information under one logical set of causes. The data are, however, consistent with the hypothesis put forward recently by FELL (1967) that those conditions which cause activation of the intracellular lysosomal system favor dedifferentiation.

III. Rapid Rate of Growth as a "Cause" of Cell Dedifferentiation

The observation has been made that rapidly dividing cells do not differentiate at all (ABBOTT and HOLTZER 1966; HOLTZER et al. 1960; WHITTAKER 1963; GROBSTEIN

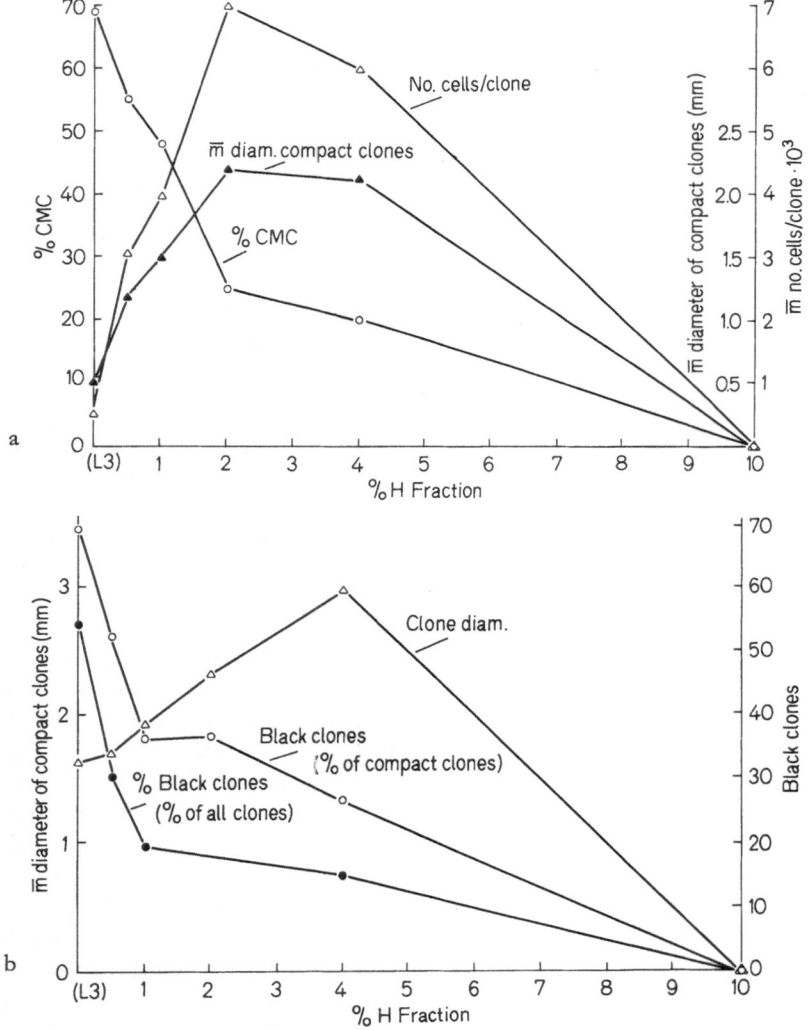

Fig. 4. Effect of increasing amounts of the H fraction of embryo extract on the growth and differentiation of chick embryo differentiated cell clones. a) Chondrocytes. b) Pigmented retina cells

1959), or differentiate poorly if they do. While this correlation appears to hold, it does not necessarily follow that rapid growth *per se* "causes" cellular dedifferentiation, or that rapid cell growth and division are incompatible with synthesis of cell specific products by differentiated cells. The following pieces of evidence are offered to support the conclusion that, given conditions which protect the cells against adverse environmental conditions (lysosomal activation?), rapidly dividing cells can and do express most of their normal differentiated properties. However, under conditions of normal clonal and mass monolayer growth cell differentiation is certainly quantitatively greater (and perhaps qualitatively different, although that remains to be proven; see NAMEROFF and HOLTZER 1967) when the cells have a relatively long generation time.

Figure 4a demonstrates that average cell growth rate or "mean generation time" is not the only factor determining whether cells are able to differentiate *in vitro*. For chick cartilage cells, both 1% and 6% EE give a generation time of approximately 23 hours. Yet, the percentage of differentiating clones in the lower concentration of EE is more than three times as high as in the higher EE concentration. Clearly more EE in the medium prevents differentiation of chick cartilage cells independently of or in addition to its effect on the mean cellular growth rate. Similar results are recorded in Fig. 4b, for pigmented retina cells.

This same point is illustrated in Fig. 5. Clones were grown either in EE or in high concentrations of FCS to obtain equivalent generation times. Rapid growth in media containing high serum concentrations usually allowed good differentiation. Certainly FCS did not inhibit it. A very low amount of EE on the other hand, promoted rapid

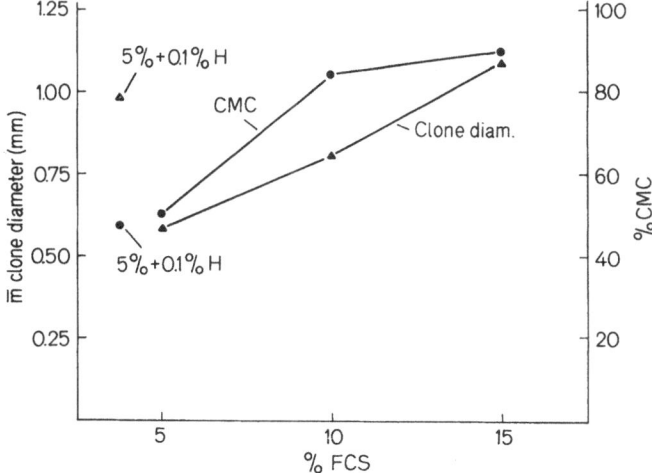

Fig. 5. Differentiation and growth of chick embryo chondrocytes with increasing amounts of fetal calf serum (FCS) added to medium. The points to the left of the curves represent cultures grown with an addition of 0.1% of the H fraction of embryo extract (EE) in 5% FCS containing medium. Triangles = clone diameter; circles = % cartilage making clones. Although the growth of the cells in the 0.1% H has been increased considerably, they show a differentiation percentage almost equal to the cultures with 5% FCS alone. Cells growing this fast in media containing only FCS have close to double the % differentiated colonies. Thus the EE contains some factors detrimental to differentiation separate from those which promote rapid growth

growth of the cells but did not stimulate or inhibit expression in these cells. Thus, in the presence of EE, equivalent large clone size does not create all the conditions necessary for maximal expression of differentiation.

One conclusion which could be drawn from these experiments is that the stability of differentiation can be influenced by some agents in EE which may also promote growth; on the other hand, rapid growth *per se* is not necessarily incompatible with expression of differentiation, but may make the cells more susceptible to substances or conditions inhibiting differentiation.

IV. Rapid DNA Synthesis:
Incompatible with Good Differentiation?

It has been proposed many times in the literature that cell growth, or more exactly, rapid DNA synthesis appears to be incompatible with simultaneous expression of maximum differentation in several cell types (ABBOTT and HOLTZER 1966; RUTTER et al. 1967; WHITTAKER 1963; EPHRUSSI and TEMIN 1960). There appears to be good evidence for this in the case of skeletal muscle (STOCKDALE et al. 1964), pancreas (BERNFIELD and GROBSTEIN 1967) and mammary tissue (STOCKDALE, JUERGENS, and TOPPER 1966), and presumptive evidence in nerve and some other glandular tissues. However, as has been shown recently by RUTTER and WESSELLS (RUTTER et al. 1967a, b, c; WESSELLS 1967), the rapidly dividing stem cell population of even some of the glandular tissues (pancreas) does possess all or most of the specialized products and synthetic machinery of the fully differentiated tissue, but in much smaller amounts. The question then becomes one not so much of "stability of the differentiated state", but control of amount of expression of differentiation under different conditions.

We have recently reported a series of experiments bearing on this question (CAHN and LASHER 1967). The experiments were designed to investigate whether cartilage cells can, during a $1-2$ hour pulse period, simultaneously synthesize DNA *and* large amounts of a sulfated mucopolysaccharide (MPS) similar, or identical to normal chrondroitin sulfate. The data are summarized in Table 6. They clearly demonstrate the presence of doubly labelled cells in cultures incubated with both ^3H thymidine and $^{35}SO_4$ for a 2 hour pulse. *All the cells synthesizing DNA* were also surrounded by considerable to enormous amounts of ^{35}S-containing material representing newly synthesized cartilage matrix. The reverse however was not true: not all the actively differentating cells (synthesizing sulphated MPS) were incorporating thymidine into DNA during the pulse period.

At least two objections can be raised to the conclusions derived from the above data. First, demonstration of nuclear DNA synthesis does not prove that the cells are dividing rapidly. One could argue that the cells may be synthesizing "metabolic" DNA, and/or that the DNA synthesis observed is terminal, prior to complete cessation of cell division, making such cells tetraploid or at least containing the 4n amount of nucleic acid. Since the populations dealt with in the experiments cited above were in the log phase of growth, and since the percent of cells labelled with ^3H-thymidine during a 2 hour pulse correlates well with the known duration of the S-phase and total cell cycle times for tissue cultured chick cells, it is likely that the majority of the nuclear DNA synthesized represents normal nuclear chromosomal

Table 6. *Simultaneous incorporation of ^3H-Thymidine and $^{35}SO_4^{--}$ by growing differentiated chondrocyte cell clones*

Type of clone	Area cells labelled over:	Enzyme or Drug Treatment	Corrected grain counts (over cytoplasm and matrix only)	Number of cells counted	Significance (p)
1. CMC	N+C+M	none	121.5 ± 35.2	30	1: 2 (p <0.001)
2. CMC	C+M only	none	83.3 ± 17.6	30	1: 9 (p <0.001)
3. CMC	N+C+M	+ Hyase	23.0 ± 16.5	30	2:10 (p <0.001)
4. CMC	C+M only	+ Hyase	11.4 ± 12.4	30	
5. CMC	N+C+M	DNAse + Hyase	18.0 ± 9.7	30	
6. CMC	C+M only	DNAse + Hyase	16.4 ± 5.8	30	
7. CMC	N+C+M	Cycloheximide	9.4 ± 2.8	30	
8. CMC	C+M only	Cycloheximide	8.2 ± 2.8	30	
9. FB	N+C+M	none	49.2 ± 20.0	45	
10. FB	C+M	none	45.2 ± 18.2	45	
11. FB	N+C+M	+ Hyase	14.8 ± 11.5	30	
12. FB	C+M only	+ Hyase	25.0 ± 11.1	30	
13. Control plates (blank background)	no label		8.0 ± 2.8	30	

Cells incubated for 2 hours in presence of both labelled compounds.

N+C+M = cells which were labelled both in nuclei, cytoplasm and over extracellular matrix.

C+M only = cells which were not labelled over the nuclei.

Counts were corrected for DNA counts in cytoplasm (see CAHN and LASHER 1967).

replication prior to cell division. However, quantitative analysis of pulse-chase experiments will have to be done to rule out turnover of DNA. Second, rapidly dividing cells *in vitro* may not be synthesizing the exact products they do *in vivo*, or at least nowhere near as much (see NAMEROFF and HOLTZER 1967). This argument has been extended to point out that the cell may *never* express exactly the same type of cytodifferentiation *in vitro* as *in vivo*, or under rapidly dividing conditions as under static conditions.

I believe much of this line of argument is semantic, not substantive. Three pieces of evidence lead me to make this statement.

1. Much of differentiation may be a selective amplification of types of specialized syntheses present in many cells at much lower levels. RUTTER and his group (RUTTER et al. 1967a, b, c) have demonstrated that even socalled "non-induced" pancreas cells synthesize considerable amounts of pancreatic enzymes. It is not too far fetched to suppose that closely related cell types (fibroblasts, osteoblasts, chondroblasts) may synthesize many of the same products (collagen, sulfated MPS) but at different rates, and in different proportions, leading to different end results.

2. COON has shown that cartilage cells grown as clones synthesize chondroitin sulfates A and/or C (COON 1964; COON and MARZULLO 1967). NAMEROFF and HOLTZER (1967) recently, in a more detailed study, showed that mass cultures of dividing chondrocytes grown under conditions quite different than those employed by COON and myself, synthesized very little or no true chondroitin sulfate, but make other

poorly sulfated MPS which contain much more glucose than the normal product. Non-growing cells synthesized more highly sulfated MPS, but it was still not "identical" with the normal product. However, since it has been demonstrated repeatedly now (COON 1966; COON and CAHN 1966; CAHN, COON and CAHN 1968) that the conditions of culture employed by HOLTZER and his group (EE, crowded conditions) prevent specialized cytodifferentiation irrespective of rate of cell growth (see previous section), it seems unwise to generalize the results of HOLTZER and his collaborators to all conditions, and conclude that DNA synthesis and mitosis *per se* are incompatible with differentiated syntheses. Studies to investigate the type of MPS being synthesized in the centers and edges of differentiated clones may shed more light on the problem of control of type of MPS synthesis by DNA synthesis rate.

One could also argue that a 1−2 hour pulse is too long a period from which to call the DNA and MPS synthesis "simultaneous". Those cells which are synthesizing DNA may have stopped and be in the G2 period prior to mitosis, when MPS synthesis commences. We have no direct evidence bearing on this point although it seems unlikely in view of the already mentioned fact that the percent of labelled cells correlates well with known S and total cell cycle times, and that all cells labelled in the nucleus were synthesizing as much or more ^{35}S-labelled MPS than the unlabelled cells (CAHN and LASHER 1967).

Recently, studies on collagen synthesis seem to support the conclusion that DNA synthesis and synthesis of specialized protein are not mutually exclusive (DAVIES, PRIEST, and PRIEST 1967). These workers investigated collagen as a "specialized protein". Again, it may be argued that collagen is synthesized by many cell types, and therefore does not represent true "specialized" protein. I feel these arguments fall into the same semantic category as those I mentioned above. Differentiation, we must now realize, may not be an "all or none" phenomenon, but may simply mean that certain functions are more "derepressed" by some cells than by others. The sum of the "derepressions" would be equal to the differentiation of a cell. Certainly, under this definition, large amounts of collagen represent a differentiated product to those particular cells.

3. Another piece of evidence on this point relates to the title of this volume, for it indicates that stability and expression of differentiation are relative, not absolute phenomena. I showed several years ago (CAHN 1964b; see also CAHN 1967b) that under conditions of monolayer tissue culture, all chick embryo cells contain very similar LDH isozyme patterns, regardless of the relative amount of the different types of LDH in the tissue *in vivo* before culture. This is illustrated in Table 7. Chick embryo heart cells, whether beating or not, contain quantitatively the same types of LDH as myoblasts, cartilage cells or pigmented retina *in vitro*. A detailed analysis reveals some significant differences in relative occurrence of the five isozymes in different chick tissue cultures, but nonetheless it is highly probable that the culture environment, perhaps tissue oxygen supply, and not the rate of cell division, determines which LDH is formed. Table 8 shows that even organ cultures, in which little or no cell division is taking place, change their LDH patterns after several days in culture to resemble the patterns of the monolayer cell cultures. While the LDH patterns change drastically from their "normal" differentiated patterns, other differentiated traits of the cells remain stable: beating in heart cell monolayers, synthesis of sulfated MPS in cartilage, and synthesis of melanin containing melanosomes in pigmented retina cells.

Table 7. *Lactate dehydrogenase patterns from differentiated and "dedifferentiated" chick embryo cell clones and monolayer cultures*

Type of cell or culture	% H-LDH Subunits	
	Before culture	After culture
1. Cardiac myoblasts (beating monolayer)	86 ± 3 ($N = 40$)	3.2 (3.0, 3.4)
2. Cardiac myoblasts (nonbeating monolayer)	—	9.8 (9.0, 10.5)
3. Clones derived from isolated beating cardiac myoblasts	—	8.0 (7.0, 9.0)
4. Clones derived from primary platings of mixtures of cardiac myoblasts + fibroblasts	—	8.0 ± 3.2 ($N = 61$)
5. Differentiated pigmented retina clones (black)	—	0 (0, 0)
6. "Dedifferentiated" pigmented retina clones (no pigment)	—	0 (0, 0)
7. Differentiated limb cartilage clones	16 (13—20)	2.8 ± 0.2 ($N = 12$)
8. "Dedifferentiated" limb cartilage clones	—	2.2 ± 0.7 ($N = 50$)
9. Skeletal myoblast monolayers	53 (49 — 59)	9.0 (7.9 — 10 5)

Where sufficient data were available a standard deviation was calculated. Otherwise a range, or in a few cases, the actual experimental values are given in parentheses. Where a standard deviation is given, N equals the number of experimental samples. The value given for cardiac myoblasts before culture was used as the value for comparison for entries 1—4. For monolayer mass cultures (entries 1,2,9) the values given are after 6 days in culture. For the clones (entries 3—8) the values are for clones of approximately 10^4 cells from 2—4 weeks after clonal passage. Results are expressed as the percent of total LDH activity represented by the H-LDH (electrophoretically fastest moving form) subunit.

Table 8. *Comparison between rates of change of lactate dehydrogenase patterns of chick embryo heart cells cultured in monolayer culture or in gyrorotatory flask culture*

Days in culture	% H- Subunits	
	Organ culture	Monolayer culture
0	86	86
1	65	76
2	65	43
4	45	17
7	—	15

Results are expressed as the percent of total LDH activity represented by the H-LDH (electrophoretically fastest moving form) subunit.

Thus I feel the stability of the expression of the differentiated state depends on the stability of the environment, while the stability of the epigenotype may rest upon as yet undiscovered information storage and retrieval mechanisms in the nucleus and/or cytoplasm of the cell.

The last two sections of this paper will deal with these two aspects in a more analytical sense. 1. At what chemical level does the environment interact with and control expression of epiphenotype? 2. What and where in the cell are the components responsible for day to day stability of the epigenotype? Are they in the chromosomes, in the nucleus or the cytoplasm, or some combination thereof?

V. Biochemical Analysis of Cellular "Dedifferentiation"

Cells may fail to synthesize their specific products for one or more of the following reasons:

1. The enzymes or structural molecules needed are synthesized at the normal rate, but leak out of the cell or are washed away into the medium too rapidly to be able to provide normal cytodifferentiation.

2. The enzymes or other macromolecules needed are synthesized at the normal rate but are degraded much more rapidly than normally, preventing normal cytodifferentiation (SCHIMKE 1964, SCHIMKE, SWEENY, and BERLIN 1965).

3. The products are not synthesized or are synthesized at a greatly reduced rate because of some defect in the mechanism of protein synthesis at a point in the chain between DNA and the ribosomal synthetic system. Here there are several possibilities:

a) informational RNA molecules are not transcribed from the DNA;

b) such m-RNA molecules are transcribed but never leave the nucleus;

c) m-RNA leaves in a non-functional state (informosomes of SPIRIN?) and cannot be activated in the cytoplasm;

d) synthesis is blocked at the polysome level by limiting m-RNA or other precursors, or by rapid breakdown, inactivation or destruction of polysomes by intracellular degradative enzymes.

Investigations of differentiation at the level of the first two possibilities have been made by MARZULLO and LASH (1967), and COON and MARZULLO (1967). Detailed studies of turnover of specialized products must await methods for analysis of the synthesis of specific protein in these systems (e.g. by specific immunochemical identification, or unequivocal identification of enzymes and other protein by polyacrylamide gel electrophoresis). We are presently studying the turnover of specific protein in both cartilage and pigmented retina cells. Here, however, I want to report briefly a line of experimentation designed to distinguish between alternatives 3a–d mentioned above. These investigations have been carried out in collaboration with Mr. MICHAEL SOLURSH in our laboratory. They rely upon the technique of DNA-RNA complementary binding ("hybridization") on membrane filters. This technique can demonstrate differences in the populations of RNA molecules synthesized in expressing and non-expressing cartilage and pigmented retina cells. The results of such experiments are summarized in Figs. 6 and 7.

Before discussing these results it is important to keep in mind several serious limitations and reservations to any conclusions which can be drawn from DNA-RNA binding experiments. In the experiments presented here, RNA labelled with tritiated uridine during a ten minute pulse was extracted from well differentiated cartilage cell clones. Such short labelling times have been chosen to reduce the relative amount of label in ribosomal RNA, which may be synthesized somewhat more slowly than the supposed "messenger" RNA. The curves in Fig. 6 represent the amount of the labelled RNA, isolated from *whole cells*, which binds to chick embryo DNA on the filters in the presence of increasing amounts of "cold" RNA from various sources. The usual assumption in interpreting these "competition curves" is that any RNA which binds to DNA under the conditions of the assay does so because of its complementarity to the DNA, thus justifying the assumption that it is the "messenger" RNA. Several facts confound this interpretation, however. One is (HARRIS 1963;

PENMAN, VESCO, and PENMAN, 1968) that the vast majority of the RNA labelled using short pulses of ^{3}H-uridine may never leave the nucleus, but turns over there very rapidly. Although some of this may represent "messenger" RNA, all of it binds

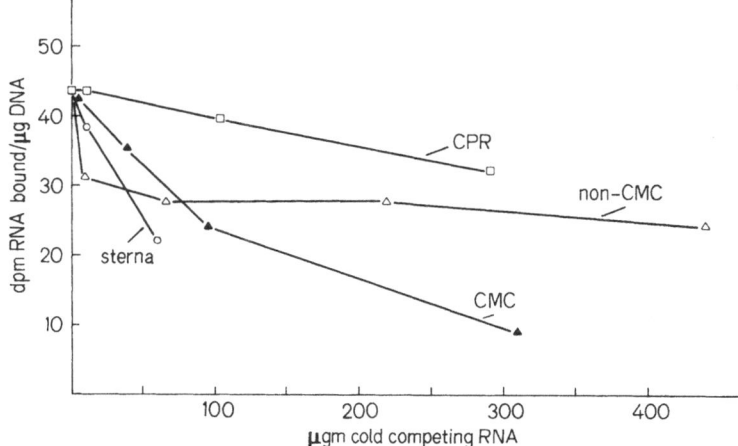

Fig. 6. Comparison of ability of RNA from various sources to interfere with the ability of ^{3}H-uridine labelled cultured cartilage making cell RNA to bind specifically to chick embryo DNA. CPR = chick pigmented retina cell clone RNA derived from well pigmented clones. CMC = RNA derived from picked 2nd passage differentiated cartilage making clones. non-CMC = RNA derived from fibroblast like cells derived originally from picked CMC clones and grown in media which prevent expression of differentiation. Sterna = RNA derived from 14 day whole chick embryo sterna. 7.4 μg of chick embryo DNA were bound to each filter (5.8 mm diameter, Schleicher & Schuell type B6). 10.2 μg of labelled RNA were in each vial, in 0.2 ml of 2 × SSC. Reactions were incubated 16 hours at 67° C

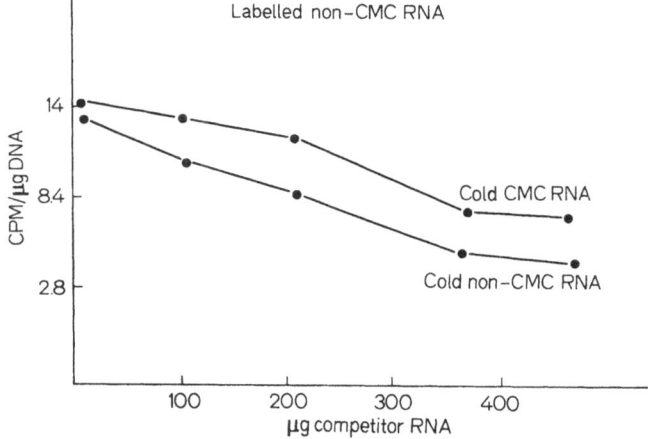

Fig. 7. Comparison of ability of differentiated cartilage cell clone RNA and dedifferentiated cartilage cell RNA to interfere with the binding of ^{3}H-uridine labelled *dedifferentiated* cartilage cell RNA to chick embryo DNA. Conditions as in Fig. 6 except that there were 30 μg of labelled non-CMC RNA per reaction mixture. Total volume 0.2 ml, 7.1 μg of DNA per 5.8 mm diameter S & S B6 filter in 2 × SSC, 16 hours, 67° C. Non-CMC RNA came from mass cultures of cartilage making cells were not differentiated

very tightly to the DNA under the usual annealing procedures (BIRNBAUM, PÉNE, and DARNELL 1967). Secondly, it appears from experiments of BRITTEN and KOHNE (1968) that the fidelity of the complementarity need not be perfect to allow DNA-RNA binding. According to these workers DNA exists in the cell in multiple copies, so that the RNA-DNA binding could be due to binding of RNA only to these "redundant" DNAs. The authors further argue that complementary binding could not, on kinetic grounds, be expected to occur to any great extent with the unique DNA sequences. Beside this, glycogen, and other polysaccharides found as contaminants in RNA preparations, may interfere with DNA-RNA binding. Finally, the quantitative aspects of this binding can be manipulated by the exact conditions of the experiment, e. g. temperature, salt concentrations, etc. Differences between similar RNA samples are magnified at lower salt concentrations, higher temperatures, and higher RNA/DNA ratios. All these facts call for extreme caution in interpretation of the results presented in Figs. 6 and 7.

Taken at face value, however, the results certainly indicate that the DNA-RNA binding technique is capable of detecting differences between rapidly labelled RNA from different cell types. All the RNAs were isolated simultaneously under identical procedures and freed of MPS and glycogen as far as possible. Sternal cartilage RNA *ex embryo* was not completely freed of MPS and only three valid points could be obtained in initial experiments before the MPS began to interfere with the binding. More recent experiments with purer RNA preparations confirm and extend all the data. The following tentative conclusions may be drawn from the experiments.

1. Cultured pigmented retina (PR) cell clones synthesize a rapidly labelled RNA (RL-RNA) which is quite different from RL-RNA isolated from clones of sternal cartilage cells (SC). Actually the differences are much greater than could be accounted for on the assumption that PR and SC cells differ in the activation of as many as 1000 different sets of genes, and that RL-RNA represents the "messenger" RNA transcribed from these genes. Accepting BRITTEN and KOHNE's hypothesis that for the most part only the redundant DNA species are important in DNA-RNA binding, there must be multiple copies of RNA which are unique to each cell type, and which derive from the redundant segments of the DNA. These RNAs and DNA's therefore must account for the bulk of the observed binding. Thus the redundant DNA's may be concerned with cell epigenotype stability (LASHER and CAHN, 1968).

2. The RL-RNA patterns of "dedifferentiated" cartilage cells are markedly different from those of differentiated cartilage cells. One possible explanation of this result is that dedifferentiated cartilage cells do not express their differentation because they lack "messenger" or other information bearing RL-RNA for some or all of the specific proteins needed for cartilage synthesis. The early steep part of the curve in Fig. 6, representing binding of non-CMC-competitor at low concentrations could mean that dedifferentiated cartilage cell RNA contains a relatively higher percentage of the more common RNA sequences than does the cartilage making cell (CMC) RL-RNA. These relatively more abundant RNA species could be involved in "maintenance" functions or could be other RL-RNA needed due to the possibly more rapid turnover of many proteins in the non-expressing cells.

The data seem to indicate further that non-CMC-RNA is as different from CMC RL-RNA as is PR-RNA. However, this does not mean that non-CMC cells are that similar to PR cells. Experiments, using RL-PR-RNA, RL non-CMC-RNA, and other reciprocal combinations are needed for drawing any conclusions about the above alternative possibilities. One such experiment is shown in Fig. 7. Here the pulse-labelled RNA was derived from "dedifferentiated" cells which were grown from picked differentiated cartilage making clones. Under the conditions of this experiment the differences between the RL-RNA from "dedifferentiated" cartilage cells and normal cells are small but significant. But it is obvious that the dedifferentiated cells possess some species of RNA which are not synthesized, or concentrated in any great amounts by the differentiated cells. Whether this RNA represents "messengers" for new protein species, or is involved in the more rapid protein turnover of dedifferentiated cells cannot be decided at present.

The conclusions which can be drawn from the above experiments are further clouded by the observations of BIRNBAUM, PÉNE, and DARNELL (1967) that a non-specific competition between the nuclear RL-RNA and cytoplasmic RNA takes place under some conditions for DNA-RNA annealing. Their experiments were carried out at 60° however, and under the conditions used in our experiments (67−68° C) such non specific-competition is be reduced considerably. All these experiments ought to be repeated using isolated, purified, $>45\,S$ nuclear RL-RNA, and cold $>45\,S$ RNA as competitor for a control. Only then we can begin to draw any molecular conclusion from the experiments.

3. Finally, it appears from all the data available at present, that cultured CMC-RNA is very similar, but not identical, to RNA from sternal cartilage *ex embryo*. Sternal cartilage appears to possess a greater proportion of specific "cartilage" RNA species than cultured CMC clones *in vitro*. These tentative conclusions have been verified with more MPS-free RNA.

In summary, the presently available data are compatible with, but in no way prove the hypothesis that "dedifferentiation" of cells cultured *in vitro* is at least in part due to the fact that these cells no longer synthesize enough of the appropriate "messenger" or other RL-RNA species under the specific environmental conditions which favor the dedifferentiation. Some or all of the other mechanisms proposed in the section above may also be operating to prevent maximal expression of dedifferentiated traits *in vitro*, but further experimentation is necessary to assess their importance. Finally, taken together with our data on BUDR inhibition of expression and inheritance of epigenotype, the evidence suggests a role for extra DNA copies in the maintenance of cellular differentiation.

VI. Utility of Cell Hybridization and Cell Fusion as a Tool for the Study of Stability of Cellular Differentiation

The problem of the site of control of the stability of cellular differentiation can be further considered in two basic, but not exclusive, categories. 1. The nucleus is "stable" and directs differentiated cytoplasmic synthesis, being relatively independent of the cytoplasm in its control of cell specialization. 2. The "stability" rests in the cytoplasm; the cytoplasm exerts a relatively self sustaining control over the differentiated syntheses, and is primarily responsible for a continuing feedback control of the nucleus and the nuclear role in the synthetic pattern.

Classic experiments by BOVERI (1887) and SPEMANN (1928) demonstrated the importance of the specialized egg cytoplasm in determining the eventual differentiated fate of the nuclei which come to lie in those areas. Thus they showed the "totipotency" of early embryonic nuclei and their plasticity in response to the cytoplasmic environment. Later, the experiments of Briggs and King (1952); King and Briggs (1954, 1956) confirmed and extended the early experiments to conclude that nuclei from cells as late as the gastrula entoderm were totipotent. Gurdon (1962) went even further and concluded that nuclei from cells of fully differentiated ciliated gut epithelium are capable of supporting complete normal development when transplanted into enucleated egg cytoplasm, i.e. they are totipotent.

These experiments however do not give us any information about whether it is the nucleus or cytoplasm which is paramount in determining the *stability* of the final differentiated state. They merely demonstrate that the nuclei or chromatin are plastic, can respond to differing cytoplasmic environments and that in the late embryo at least the nuclei of *some* cells (totipotent "germ cells"?) possess all the information needed to allow complete normal development of an early embryo.

But we must still ask the following questions. Would a nucleus of a cartilage cell, transplanted into a pigment cell, support continued pigmented retina cell differentiation? When such a cell multiplied, would the resultant clone be a cartilage cell, a pigmented retina cell, both, or neither? Is there a hierarchy of nuclear control of differentiation, i.e. are certain types of differentiated cell nuclei dominant in combination with certain cell cytoplasms, but recessive in other combinations? How do combinations behave between different nuclei or different cytoplasms, or those from nuclei derived from different time points in the continuum of development?

These are questions of epigenetic dominance and recessiveness, and we can ask: Are there methods at hand which can help to answer such questions? What could such experiments actually tell us about the stability of differentiation?

Ephrussi has pioneered in the use of somatic cell hybridization in studies of epigenetic stability. In fact he had predicted some of the more recent lines of experimentation in even his earliest papers. Briefly, Ephrussi and his group have formed hybrid cells from rat and mouse cells which possess nuclei with most of (or nearly all) the chromosomes of both parents by allowing the cells to fuse spontaneously at a lowered temperature. The hybrid cells were then selected by a modification of the technique of Littlefield (1964) as described in detail by Davidson and Ephrussi (1965) and Ephrussi and Weiss (1965). The selective technique involves the use of cells which lack either inosinic acid pyrophosphorylase (resistant to 8-azaguanine, 8-AG) or thymidine kinase (resistant to bromodeoxyuridine, BUDR). Hybrid cells can grow in a medium containing hypoxanthine, aminopterin, and thymidine (HAT medium), in which neither parent cell can grow; the BUDR resistant parent cannot utilize exogenous thymidine because it lacks thymidine kinase, and it cannot make its own thymidine due to the presence of aminopterin; the 8-AG resistant parent lacks the enzymes needed to allow the utilization of hypoxanthine which is required for its growth. Hybrid cells produce enzymes which release both these blocks. In some of the hybridizations, Ephrussi used cells with only the BUDR resistant marker and a normal cell as the other parent (a half selected cross). The hybrids grew in the HAT medium and overgrew the normal parent.

Using this type of selection, Ephrussi and his group, and more recently Coon and his group (1967), have carried out hybridizations between various types of differentiated cells and "fibroblasts" or other relatively undifferentiated tumor cells or permanent cell lines. In all cases studied so far by EPHRUSSI, the overt "differentiated" traits (pigment production by melanomas, teratoma production by teratoma cell lines) were extinguished in the hybrids, while other genes of both parents (lactate dehydrogenase or β-glucuronidase) were still functional. In the case of cells making collagen or hyaluronic acid (GREEN et al. 1966), the hybrid cells had levels of synthesis of these substances almost exactly intermediate between those of the two parental cell types. The conclusion from these experiments was that expression of differentiated traits is controlled by repression of the other genes; and that in the hybrids the repressors from the non-differentiated cell parent are actively synthesized and are "turning off" the differentiated traits of the other cell line. Since neither EPHRUSSI nor COON has yet performed hybridizations between two overtly differentiated cell types, the validity of this interpretation is still open to question. We know that BUDR containing cells can be "cured" of their differentiation. Although there may be no BUDR in the BUDR resistent cells, nor is there any in the selective medium, there is a good chance that the progenitor cells which gave rise to the cell lines in question were incorporating at least some BUDR into their genomes. Whether the presence of the drugs in the medium at the time of genesis of the BUDR resistant parent, the absence of thymidine synthesis, or other parameters affect differentiation is not clear from their experiments, although presumably the differentiated cells could still express their differentiation in the HAT medium. This they have not yet demonstrated, however, and our own work suggests that HAT medium may inhibit the differentiation of some cells. That the mutual repression of differentiated traits is not absolute, is demonstrated by their experiments with collagen and hyaluronate.

Although the experiments using the half selected cross (BUDR resistant by normal cell) are a real beginning towards a genetic analysis of differentiation and its control, we feel that it is important to carry out such hybridizations with cells which have never been exposed to BUDR (for reasons: compare this article, section II E). We are engaged in such hybridizations, using Sendai virus to raise the spontaneous fusion rate, and have worked out conditions for the isolation of dikaryons from pigmented retina and cartilage cells of chick and quail (CAHN, CAHN and LASHER 1967; see also Fig. 8). Furthermore we are extending such studies to differentiated mammalian cells. At present the fused cells give rise only to very abortive clones in most instances. However, the percentage of viable colonies formed is only 1 in 10^4 in the system being used by EPHRUSSI and COON, and it is therefore no surprise to us that to date we have not found any true hybrid colonies with good viability. Work continues on methods to improve the hybridization and to develop a selection procedure not dependent on BUDR exposure (methods see: CAHN, CAHN and OKADA 1968).

In a recent paper, WEISS and GREEN (1967) showed that certain hybrids can spontaneously eliminate chromosomes, thus allowing chromosome mapping of certain functions of the hybrid cells. These techniques could possibly be applied to the study of differentiated functions of the cells as well.

Using cell fusions produced by methods developed in detail by OKADA (see e.g. OKADA 1962a) HARRIS et al. (1965, 1966) we have used a somewhat different approach. HARRIS also used the Sendai Virus (Haemagglutinating Virus of Japan) to fuse together various types of cells, both differentiated and relatively unspecialized. The net result of these experiments be summarized as follows: although the nucleic acid synthesis of both parent cell types seems to continue, even to the extent that RNA and DNA synthesis are

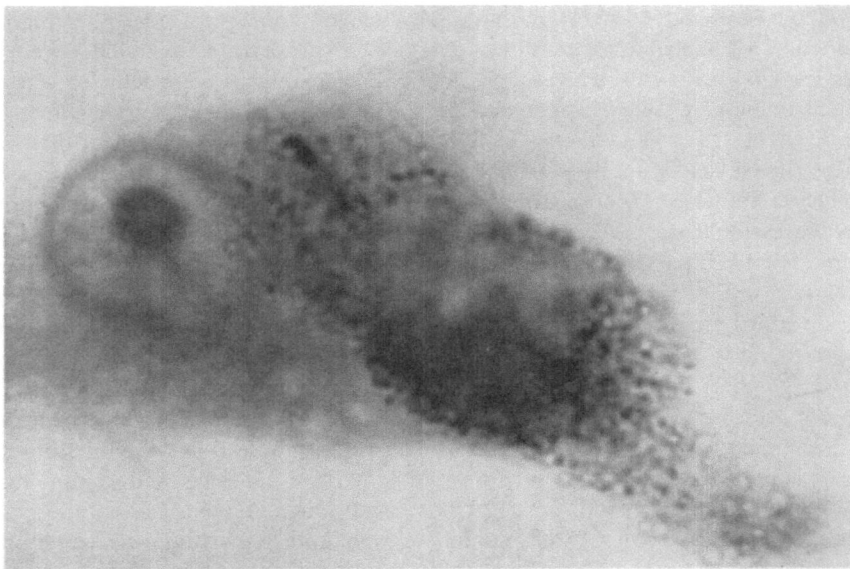

Fig. 8. Early fusion of chick pigmented retina cell with quail cartilage cell. Magnification approximately 1100 × on plate. Chick pigmented retina cells, attached to petri dishes, were fused with quail cartilage cells by means of Sendai Virus induced membrane fusion. Cells on plate were treated in cold with approximately 4000 HAU/ml u.v. inactivated Sendai virus, incubated with 2 × 10⁶ quail cartilage cells/plate, washed, and incubated at 37° C. Individual ringed pigmented retina cells were then observed to see if they had fused with a cartilage cell. The cell in this photograph went on to completely intermingle cytoplasms, so that pigment granules were fairly evenly dispersed. The beginnings of the process can already be observed in the photograph shown here. Such fused cells, however, gave very few spontaneous colony formers, most of which, without selective techniques involving BUDR, have been abortive and very small to date

initiated in an erythrocyte nucleus when introduced into a Hela cell, there is no proof that any of the differentiated properties of the cells are maintained in the hybrids. Harris (personal communication) even has doubts that the RNA made by the reactivated red cell nucleus can function in the Hela cell cytoplasm to enable production of erythrocyte specific products (haemoglobin). Watkins and Grace (1967) have demonstrated, however, that one can get some clones of virus fused cells which retain the ability to make both parental cell surface antigens, although many clones seem to have lost the ability to make one or the other. Most of the studies of the Harris group, however, have been made with heterokaryons, in which the contribution of the initial cytoplasm is very large compared to its role in the syntheses of the cells in the hybrid clones of Ephrussi. Thus molecular interpretations at this stage are impossible.

Several groups are working on efforts to combine the best parts of both methods. Coon (personal communication) is utilizing the half selective technique in back-crosses to normal cells to enable the separation of the contribution of the initial drug resistance markers from the possibility of a true extinction of differentiated traits. In this technique a hybrid cell formed between a drug resistant undifferentiated cell and a normal differentiated cell is backcrossed with the differentiated cell. The results of such interesting genetic analyses should be most exciting.

We in our laboratory are taking a somewhat different approach. Still working with fusions induced by killed Sendai virus, and utilizing methods worked out by OKADA which allow a highly efficient production of dikaryons of known phenotypes (CAHN et al. 1967a, b) we are fusing pigmented retina, cartilage, and other differentiated cells without the use of a selective medium. The selection comes by working with cell types and experimental conditions which reduce to a minimum all unfused cells. It has long been known (OKADA, personal communication) that mitoses and clone formation in polykaryons (more than two nuclei) are rare, while dikaryons will go on to give rise to clones with relatively high efficiency. The results of this type of analysis will enable a useful comparison to those studies involving the use of drugs and selective media.

Still another approach which we are taking will allow the analysis of fusions in which only one set of parental chromosomes is present. EPHRUSSI and COON both analyze tetraploid clones. In contrast, our method involving the treatment of the cells with Cytochalasin B (CARTER 1967), may allow us to fuse enucleated cells of one epiphenotype with normal cells of another. A somewhat different approach, the X-irradiation or UV microbeam irradiation of one cell type to prevent multiplication of that cell nucleus has been tried in several laboratories (HARRIS, personal communication; our own laboratory, unpublished), but to date the results have not been conclusive. X-irradiation or UV irradiation may damage more than just the nuclear DNA, and in any event seems to interfere with colony production. UV microbeam or laser inactivation of the nucleus of one cell type is feasible, but tedious, and has not yet yielded viable clones of hybrid cells containing the nucleus of only one parent.

So much for the possibilities. What has been demonstrated clearly recently (see Fig. 8) is that cell fusion and cell hybridization, be it with virus or drug-selection, can be used fruitfully to study problems of cellular differentiation. The answers to the other questions posed in the beginning of this section will have to wait until the full potential of these powerful tools is realized.

References

ABBOTT, J., and H. HOLTZER: The loss of phenotypic traits by differentiated cells. III. The reversible behaviour of chondrocytes in primary cultures. J. Cell Biol. 28, 473—487 (1966).
— — The loss of phenotypic traits by differentiated cells. V. The effect of 5-BUdR on cloned chondrocytes. Proc. nat. Acad. Sci. (Wash.) 59, 1144—1151 (1968).
BERNFIELD, M., and C. GROBSTEIN: Personal communication and comments at the West Coast Regional Meerings of the Society for Developmental Biology, Asilomar (1967).
BERSON, S. A., and R. S. YALOW: Quantitative aspects of the reaction between insulin and insulin-binding antibody. J. clin. Invest. 38, 1996—2016 (1959a).
— — Species specificity of human anti-beef, pork insulin serum. J. clin. Invest. 38, 2017—2025 (1959b).
BIRNBAUM, H. C., J. J. PÉNE, and J. E. DARNELL: Studies on hela cell nuclear DNA-like RNA by RNA-DNA hybridization. Proc. nat. Acad. Sci. (Wash.) 58, 320—327 (1967).
BOVERI, TH.: Über die Befruchtung des Eies von Ascaris megalocephala. S.-B. Ges. Morph. Phys. Bd. 3, München (1887).
BRIGGS, R., and T. J. KING: Transplantation of living nuclei from blastula cells into enucleated frogs' eggs. Proc. nat. Acad. Sci. (Wash.) 38, 455—473 (1952).

BRITTEN, R. J., and D. E. KOHNE: Repeated sequences for DNA. Science **161**, 529—540 (1968.)

CAHN, R. D.: Maintenance of beating and dissociation of biochemical and functional differentiation of beating and dissociation of biochemical and functional differentiation in clones of chicken embryo heart cells. J. Cell Biol. **23**, 17 A (1964a).

— Developmental changes in embryonic enzyme patterns: The effect of oxidative substrates on lactic dehydrogenase in beating chick embryonic heart cell cultures. Develop. Biol. **9**, 327—346 (1964 b).

— Detergents in membrane filters. Science **155**, 195—196 (1967).

— Changes in enzyme patterns in functional heart pigmented retina and cartilage cells in culture. In: Factors influencing myocardial contractility. (Ed.: R. D. TANZ, J. KAVALER, and J. ROBERTS), pp. 293—299. New York-London: Academic Press 1967.

—, and M. B. CAHN: Heritability of cellular differentiation: Clonal growth and expression of differentiation in retinal pigment cells in vitro. Proc. nat. Acad. Sci. (Wash.) **55**, 106—114 (1966).

— —, and R. LASHER: Heritability of cellular differentiation: Studies on growth and differentiation of mononuclear cells and interspecific fusion hybrids between pigmented retina and cartilage cells. J. Cell Biol. **35**, 20 A (1967).

— —, M. SOLURSH: Heritability of differentiation *in vitro*: Control of expression at the level of synthesis of DNA-like RNA in cartilage and pigmented retina cell clones. Amer. Zool. **7**, 750 (1967).

— —, and Y. OKADA: Methods for controlled fusion of various differentiated chick and quail cells grown *in vitro*. Manuscript in preparation (1968).

—, and A. CLARK: unpublished.

—, H. C. COON, and M. B. CAHN: Growth of differentiated cells: Cell culture and clonic techniques. In: Methods in developmental biology (Ed.: F. WILT, and N. WESSELLS), pp. 493—530. New York: Thomas Y. Crowell 1968.

—, and R. LASHER: Simultaneous synthesis of DNA and specialized cellular products by differentiating cartilage cells *in vitro*. Proc. nat. Acad. Sci. (Wash.) **58**, 1131—1138 (1967).

COLEMAN, J.: Symposium presentation at VIII Internation. Congress of Embryology. Interlaken, Switzerland (1967).

—, and A. W. COLEMAN: Reversible inhibition of clonal myogenesis by 5-BUDR. J. Cell Biol. **31**, 22 A (1966).

COON, H. C.: The retention of differentiated function among clonal and subclonal progeny of precartilage and cartilage cells from chicken embryos. J. Cell Biol. **23**, 20 A (1964).

— Clonal stability and phenotypic expression of chick cartilage cells *in vitro*. Proc. nat. Acad. Sci. (Wash.) **55**, 66—73 (1966).

— Hybrid cell strain formation by virus-induced fusion of colcemid-arrested metaphases. J. Cell Biol. **35**, 27 A (1967).

—, and R. D. CAHN: Differentiation *in vitro*. Effects of sephadex fractions of embryo extract. Science **153**, 1116—1119 (1966).

—, and G. MARZULLO: Chondroitin sulfate synthesis in clonal cultures of embryonic chick cartilage. Abstract. Seventh Internat. Congr. of Biochem. (1967).

COULOMBRE, A. J.: Correlations of structural and biochemical changes in the developing retina of the chick. Amer. J. Anat. **96**, 153—190 (1955).

COX, R. P., and C. M. MACLEOD: Alkaline phosphatase content and the effects of prednisolone on mammalian cells in culture. J. genet. Physiol. **45**, 439—485 (1962).

DAVIDSON, R., and B. EPHRUSSI: A selective system for the isolation of hybrids between L cells and normal cells. Nature (Lond.) **205**, 1170—1171 (1965).

DAVIES, L. M., J. H. PRIEST, and R. E. PRIEST: Collagen synthesis by cells synchronously replicating DNA. Science **159**, 91—93 (1967).

EPHRUSSI, B., and H. M. TEMIN: Infection of chick iris epithelium with the Rous Sarcoma virus *in vitro*. Virology **11**, 547—552 (1960).

EPHRUSSI, B., and M. C. WEISS: Interspecific hybridization of somatic cells. Proc. nat. Acad. Sci. (Wash. **53**, 1040—1042 (1965).

— — Regulation of the cell cycle in mammalian cells: Inferences and speculations based on observations of interspecific somatic hybrids. *In: Control mechanisms in developmental processes* (26th Symposium of the Society for Developmental Biology). (Ed.: M. LOCKE). New York: Academic Press 1968.

FELL, H. B.: Personal communication (1967).

GREEN, H., B. EPHRUSSI, M. YASHIDA, and D. HAMERMAN: Synthesis of collagen and hyaluronic acid by fibroblast hybrids. Proc. nat. Acad. Sci. (Wash.) **55**, 41—44 (1966).

GROBSTEIN, C.: Differentiation of vertebrate cells. In The Cell (Ed.: J. BRACHET, and A. MIRSKY), Vol. **1**, pp. 437—496. New York: Academic Press 1959.

— Interaction among cells in relation to cytodifferentiation. J. exp. Zool. **157**, 121—125 (1964).

GURDON, J. B.: The developmental capacity of nuclei taken from intestinal epithelium cells of feeding tadpoles. J. Embryol. exp. Morph. **10**, 622—640 (1962).

HARRIS, H.: Nuclear ribonucleic acid. Progr. Nucl. Acid Res. **2**, 19—59 (1963).

— Behaviour of differentiated nuclei in heterokaryons of animal cells from different species. Nature (Lond.) **206**, 583—588 (1965).

— Hybrid cells from mouse and man (Lond.): a study in genetic regulation. Proc. roy. Soc. Ser. B. **166**, 358—368 (1966).

— The reactivation of the cell nucleus. J. Cell Sci. **2**, 23—32 (1967).

—, and J. F. WATKINS: Hybrid cells derived from mouse and man: Artificial heterokaryons of mammalian cells from different species. Nature (Lond.) **205**, 640—646 (1965).

— —, G. LE M. CAMPBELL, E. P. EVAMS, and C. E. FORD: Mitosis in hybrid cells derived from mouse and man. Nature (Lond.) **207**, 606—608 (1965).

— —, C. E. FORD, and G. I. SCHOEFL: Artificial Heterokaryons of animal cells from different species. J. Cell Sci. **1** 1—30 (1966).

HOLTZER, H., J. ABBOTT, J. LASH, and S. HOLTZER: The loss of phenotypic trait by differentiated cells *in vitro*. I. Dedifferentiation of cartilage cells. Proc. nat. Acad. Sci. (Wash.) **46**, 1533—1542 (1960).

KING, T. J., and R. BRIGGS: Transplantation of living nuclei of late gastrula into enucleated eggs of *Rana pipiens*. J. Embryol. exp. Morph. **2**, 73—80 (1954).

— — Serial transplantation of embryonic nuclei. Cold Spr. Harb. Symp. quant. Biol. **21**, 271—290 (1956).

KOHNE, D.: Symposium presentation at VIII International Congress of Embryology, Interlaken, Switzerland (1967).

LASHER, R., and R. CAHN: The effects of 5-BUdR on the differentiation of chondrocytes in vitro. Develop. Biol. (1968) in press.

LITTLEFIELD, J. W.: Selection of hybride from matings of fibroblasts *in vitro* and their presumed recombinants. Science **145**, 709—710 (1964).

MANASEK, F. J.: Mitosis in developing cardiac muscle. J. Cell Biol. **37**, 191—195 (1968).

MARZULLO, G., and J. W. LASH: Separation of phosphorylated and UDP derivatives of hexosamines and acetylhexosamines by TLC. Anal. Biochem. **18**, 579—582 (1967).

— — Acquisition of the chondrocytic phenotyp. *In: Morphological and biochemical aspects of cytodifferentiation*. Basel: Karger 1968.

MELNYKOVYCH, G.: Glucocorticoid-induced resistance to deoxycholate lysis in hela cells. Science **152**, 1086—1087 (1966).

NAMEROFF, M., and H. HOLTZER: Loss of phenotypic traits by differentiated cell. IV. Changes in polysaccharides produced by dividing chondrocytes. Develop. Biol. **16**, 250—281 (1967).

OKADA, Y.: Analysis of giant polynuclear cell formation caused by HVJ virus from Ehrlich's ascites tumor cells. I. Microscopic observation of giant polynuclear cell formation. Exp. Cell Res. **26**, 98—107 (1962a).

— Analysis of giant polynuclear cell formation caused by HVJ virus from Ehrlich's ascites tumor cells. III. Relationship between cell condition and fusion reaction of cell degeneration reaction. Exp. Cell Res. **26**, 119—128 (1962b).

—, and Y. HOSOKOWA: Isolation of a new variant of HVJ showing low cell fusion activity. Biken J. **4**, No. 3, 217—220 (1961).

—, and F. MURAYAMA: Multinucleated giant cell formation by fusion between cells of different strains. Biken J. **8**, 7—12 (1965).

OKADA, Y., and F. MURAYAMA: Requirement of calcium ions for the cell fusion reaction of animal cells by HVJ. Exp. Cell Res. **44**, 527—551 (1966).

— —, and K. YAMADA: Requirement of energy for the cell fusion reaction of Ehrlich ascites tumor cells by HVJ. Virology **27**, 115—130 (1966).

Okada, Y., and J. Tadokoro: Analysis of giant polynuclear cell formation caused by HVJ virus from Ehrlich's ascites tumor cells. II. Quantitative analysis of giant polynuclear cell formation. Exp. Cell Res. **26**, 108—118 (1962).

— — The distribution of cell fusion capacity among several cell strains of cells caused by HVJ. Exp. Cell Res. **32**, 417—430 (1963).

— K. Yamada, and J. Tadokoro: Effect of antiserum on the cell fusion reaction caused by HVJ. Virology **22**, 397—409 (1964).

Penman, S., C. Vesco, and M. Penman: Localization and kinetics of formation of nuclear heterodisperse RNA, cytoplasmic heterodisperse RNA and polyribosome-associated messenger RNA in HeLa cells. J. molec. Biol. **34**, 49—69 (1968).

Rutter, W. J., W. Ball, W. Bradshaw, W. R. Clark, and T. C. Sanders: Morphological and molecular analogy in cytodifferentiation. In: Secretory mechanisms of salivary glands. (Ed.: L.H. Schneyer, and C.A. Schneyer), pp. 238—253. New York: Academic Press 1967a.

— — — W. R. Clark, and T. G. Sanders: Levels of regulation in cytodifferentiation. In: Experimental Biology and Medizin, Vol. **1**, pp. 110—124. Basel-New York: S. Karger 1967c.

—, W. R. Clark, J. D. Kemp, W. S. Bradshaw, T. G. Sanders, and W. D. Ball: Multiphasic regulation in cytodifferentiation. In: Epithelial-Mesenchymal Interactions. (Ed.: Raul Fleischmajer). Baltimore: Williams & Wilkins 1967b.

Schimke, R. T.: The importance of both synthesis and degradation in the control of arginase levels in rat liver. J. biol. Chem. **239**, 3808—3817 (1964).

—, E. W. Sweeny, and C. M. Berlin: The rolse of synthesis and degradation in the control of rat liver tryptophan purrolase. J. biol. Chem. **240**, 322—333 (1965).

Spemann, H.: Die Entwicklung seitlicher und dorso-ventraler Keimhälften bei verzögerter Kernversorgung. Z. wiss. Zool. **132**, 105—134 (1928).

Spiegelman, S.: Differentiation as the controlled production of unique enzymatic patterns. Symposia. Soc. exp. Biol. **2**, 286—325 (1948).

Spirin, A. S.: On "masked" forms of messenger RNA in early embryogenesis and in other differentiating systems. *In: Current topics in developmental biology* (Ed.: A. Monroy, and A. A. Moscona). New York-London: Academic Press 1966.

Stockdale, F., K. Okazaki, M. Nameroff, and H. Holtzer: 5-Bromodeoxyuridine: Effect on myogenesis *in vitro*. Science **146**, 533—535 (1964).

—, W. G. Juergens, and Y. J. Topper: A histological and biochemical study of hormone dependent differentiation of mammary gland tissue *in vitro*. Develop. Biol. **13**, 266—281 (1966).

Watkins, J. F., and D. M. Grace: Studies on the surface antigens of interspecific mammalian cell heterokaryons. J. Cell Sci. **2**, 193—204 (1967).

Weiss, M. C., and H. Green: Human-mouse cell lines containing partial complements of human chromosomes and functioning human genes. Proc. nat. Acad. Sci. (Wash.) **58'** 1104—1111 (1967).

Wessells, N.: Address to VIII Intl. Congress on Embryology, Interlaken, Switzerland (1967).

Whittaker, J. R.: Changes in morphogenesis during the dedifferentiation of chick retinal pigment cells in cell culture. Develop. Biol. **8**, 99—127 (1963).

— Loss of melanotic phenotype *in vitro* by differentiated retinal pigment cells: Demonstration of mechanisms involved. Develop. Biol. **15**, 553—574 (1967).

Dedifferentiation and Metaplasia in Vertebrate and Invertebrate Regeneration*

Elizabeth D. Hay

Harvard Medical School, Department of Anatomy, Boston, Mass. 02115

I. Introduction

"Is it necessary in a theory of differentiation to assume that the differentiated state is irreversible?" In our present era, the answer to this question seemingly is "no". Studies of genetic control mechanisms in bacteria make it reasonable to believe that most of the nuclear DNA is in the form of operons and regulatory genes which can be directly or indirectly influenced by the intracellular and extracellular environment. Stable patterns of gene activity responding to reversible feedback controls could be established without deletion of any part of the genetic apparatus (Jacob and Monod 1961). Thus, we have theoretical reasons for a negative answer to the qestion. Moreover, experimental evidence has accumulated in recent years suggesting that somatic plant cells (Braun, this volume) and animal nuclei (Gurdon 1964) can in some cases revert to embryonic morphology and regain total developmental capacity even though they were fully differentiated.

In the case of the regenerating systems that are the subject of the present chapter, there has been for a long time ample evidence that differentiated cells, for example myocytes and chondrocytes, are capable of throwing off vestiges of what we call differentiation and of proliferating in a manner reminiscent of embryonic cells (Butler 1933; Thornton 1938). Weiss (1939) suggested that these transformations be termed modulations, so firmly entrenched was the idea that "determination" or limitation of developmental potency parallels or precedes cytodifferentiation. Butler (1933) used the term *dedifferentiation*, to describe the cartilage alteration that occurs in regeneration, a process in which cartilage matrix is dissolved and the released cells acquire "the general characteristics of mesenchyme cells". In using this terminology, he subscribed to the opinion, now rather widely accepted, that a differentiated cell be defined by the collective specializations of structure and chemistry that equip it for a particular array of functional activities. Loss of recognizable specializations is dedifferentiation. To insist today that the term dedifferentiation also means broadening of the "developmental potency" of the cell invites only terminological mayhem (Hay 1966a, b). We would further refine Butler's definition of dedifferentiation by requiring that it connote acquisition of the "embryonic" features specifically associated with cell proliferation, such as cytoplasmic basophilia, RNA and DNA replication, and an increased nucleocytoplasmic ratio (Hay 1966b). In the

* Supported by Grant HD-00143 from the United States Public Health Service.

descriptive sections to follow, we shall see that these two processes, loss of somatic cell specializations and acquisition of specializations for cell division occur together in regenerating systems. We shall then consider briefly the available evidence for and against *metaplasia* in regeneration, before returning to discuss further the significance of the association between dedifferentiation and cell proliferation.

II. Cytological Aspects of Dedifferentiation in Vertebrate Regeneration

More than any other vertebrate organ, the amphibian limb has over the centuries captivated the attention of the embryologist interested in regeneration (Spallanzani 1789; Fraisse 1885; Weiss 1939). It was argued for a time that reserve cells gave rise to the regeneration bud or blastema that forms after amputation of the limb. Even in recent years, some review articles have embraced the idea that fibroblasts alone form the regeneration blastema in adult newts (Nicholas 1955; Schmidt 1966), a theory rendered untenable by Chalkley's exacting analysis of mitotic indices in regenerating newt limbs (Chalkley 1954). What is sometimes lost sight of is the fact that the blastema does not suddenly appear as an outgrowth at the tip of the amputated limb, but rather that the entire limb stump for a considerable distance proximal to the amputation level is involved in the process of creating the regeneration cells. In the case of the adult newt, it may take days for the epithelium to seal off the cut surface of the limb because of interference by the projecting bony skeleton. During this period, the traumatized muscle first begins to break up in the distal regions and then the process of dedifferentiation extends proximally (Thornton 1938). Using tritiated thymidine to detect sites of DNA synthesis, Hay and Fischman (1961) showed that the nuclei of the dedifferentiating muscle fibers are beginning to make DNA. Thus, for the most part the cells are not dying even though there is massive degeneration and disappearance of myofibrillar material. Eighteen days after amputation, the distal areas of the regenerating limb are occupied largely by fibroblast-like cells in the case of the adult newt. Some of these cells must have derived from the dedifferentiating muscle since their nuclei were obviously viable (Chalkley 1954; Hay and Fischman 1961). The regeneration cells or so-called blastema cells surround the bony skeleton which has itself undergone some dissolution (arrow, Fig. 1 A).

The processes of dedifferentiation that lead to formation of the blastema are more impressive in the larval salamander because the skeleton is not yet ossified. Observations of cartilage dedifferentiation in living appendages (Hay 1962) emphasize dramatically the important fact that the regeneration blastema arises backwards at the expense of the formed tissues of the stump. Almost the entire cartilaginous skele-

Fig. 1. Autoradiographs of longitudinal sections from regenerating limb of adult *Triturus viridescens* fixed in Bouin's fluid 18 days after amputation. Tritiated thymidine (5 μc) was administered intraperitoneally 3—7 hours prior to fixation (Hay and Fischman 1961). The distal portion of the skeleton has undergone considerable dissolution (arrow, A). Dedifferentiating muscle is well labelled by the isotope, indicating onset of DNA synthesis in the nuclei of the dissociating fibers (B). In the region X, mononucleate cells derived from muscle are in intermediate stages of dedifferentiation (see Fig. 2). The region labelled, blastema cells, was previously occupied by muscle cells. Schwann cells, fibroblasts and some of the skeletal cells (arrow 1 B) also incorporate tritiated thymidine and contribute to the regeneration blastema. Paraffin sections, stained with Mallory trichrome. A, 100 × ; B, 500 ×

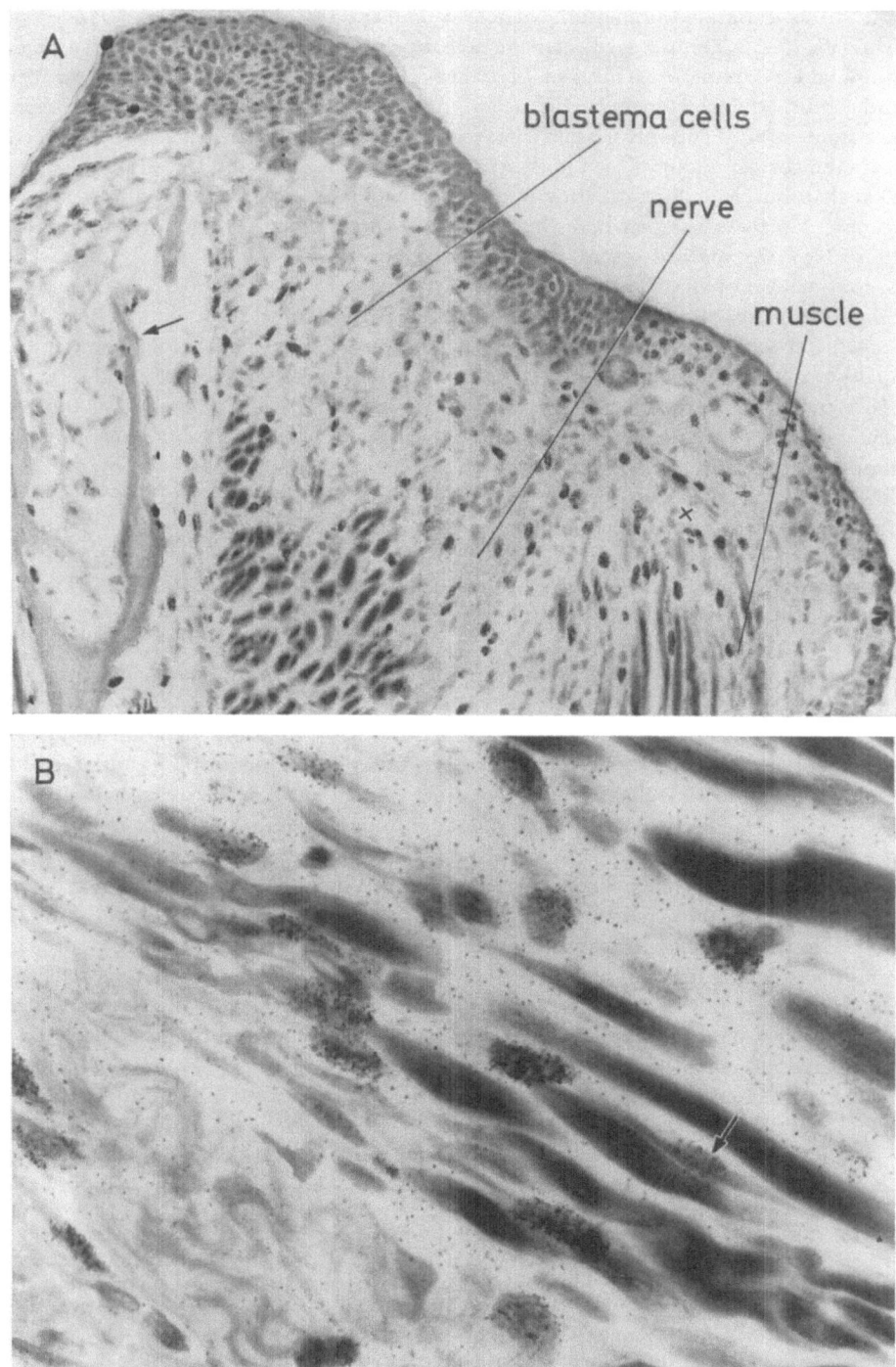

Fig. 1

ton in the amputated limb stump tends to undergo dissolution in larval *Ambystoma* (BUTLER 1933). The released cartilage cells begin to proliferate and they assume the simplified morphology typical of blastema cells. It seems likely that Schwann cells and certain other cell types also dissociate and contribute to the regeneration blastema, but interestingly enough vessels, nerves and epidermis seem to grow out directly from the corresponding tissue in the stump in both the larva and the adult. Also, it is clear that muscle dedifferentiation in the most proximal regions of the stump is never complete in the sense that here the cells retain some of their myofibrils (THORNTON 1938; NORMAN and SCHMIDT, 1967) and develop only muscle-like polysomes, that is, long chains of ribosomes in helical array (DECK, 1968).

Electron microscopic studies of dedifferentiation in the regenerating limb have focussed attention on muscle because its transformation is so dramatic. These multinucleate fibers break up into individual, mononucleate cells that take on the typical fine structure of blastema cells. Myofibrils disappear completely from the cytoplasm in all but the more proximal regions of the regenerate. Anucleate fragments of muscle disappear mysteriously. There seems to be little phagocytic activity in the area (SINGER, WEINBERG, and SIDMAN 1955). Acid phosphatase levels are high (SCHMIDT and WEARY 1963), but morphological configurations resembling lysosomes are not particularly obvious when the dedifferentiating cells are viewed in the electron microscope. There can be little doubt that the cells depicted in Fig. 2 and 3 and in our earlier work are dedifferentiating muscle cells and not phagocytes that have picked up degenerating myofibrillar material as was suggested by SINGER (1965). The myofibrils are longitudinally arranged in the cytoplasm and the cells do not have fine structural features characteristic of phagocytes, such as numerous vesicles and vacuoles. The cytoplasm, moreover, is becoming intensely basophilic. In the adult newt, the dedifferentiating muscle cells may acquire numerous segments of granular endoplasmic reticulum as well as free cytoplasmic ribosomes. The blastema cells in this case are characterized by relatively abundant granular reticulum, whereas in the larval salamander the regeneration cells have but a small to moderate amount of endoplasmic reticulum, a difference which has led to disagreement among workers in the field as to the exact state of "differentiation" of the blastema cells (SALPETER and SINGER 1960; HAY 1962). In any case, the electron micrographs make it difficult to doubt the thesis that dedifferentiating muscle cells can transform into mesenchymatous cells and that these fibroblast-like cells originating from many of the formed tissues of the stump are capable of redifferentiating into muscle, cartilage and the various connective tissues that characterize the reformed limb.

There are of course a number of other organs or organ-systems that can regenerate remarkably well in the amphibian (see ROSE 1964). Dedifferentiation of smooth muscle and other tissues has been described during regeneration of the intestine.

Fig. 2. Electron micrograph of a region of dedifferentiating muscle in the area marked X in Fig. 1 A. The fibers are breaking up into mononucleate cells and myofibrils are disappearing. Nuclei are synthesizing DNA and as they divide, the nucleocytoplasmic ratio increases. Nuclei are not included in the plane of section of fragments A, B and C. The binucleate fragment D will separate into two or more cells. The new cell membranes are formed by fusion of small vesicles seemingly derived from the sarcoplasmic reticulum. Figs. 2 and 3 are micrographs of araldite sections, stained with uranyl acetate and lead citrate, from an adult *Triturus* limb fixed 17 days after amputation through the forearm. Glutaraldehyde was used as a fixative, followed by osmium tetroxide. 5000 ×

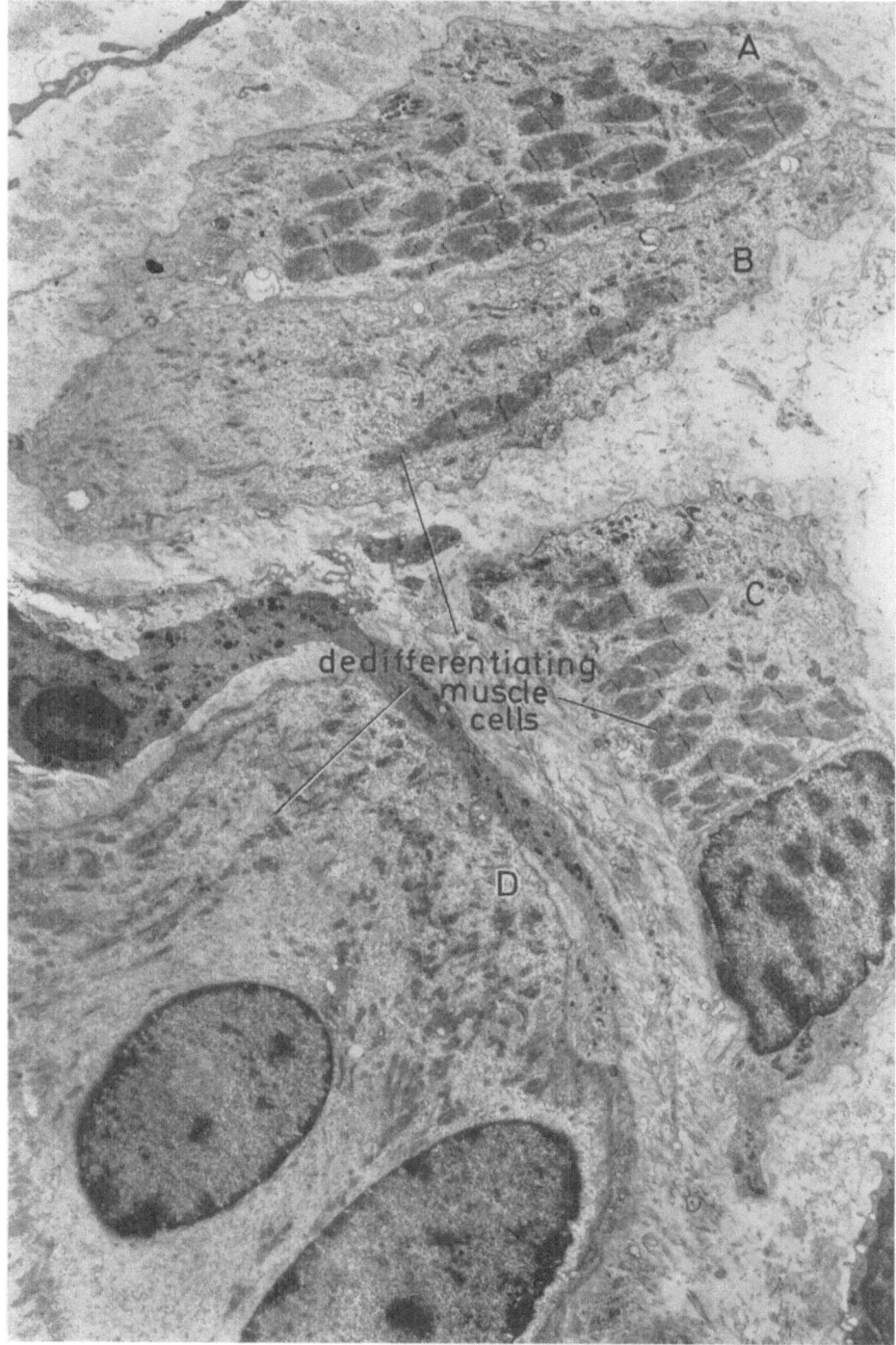

Fig. 2

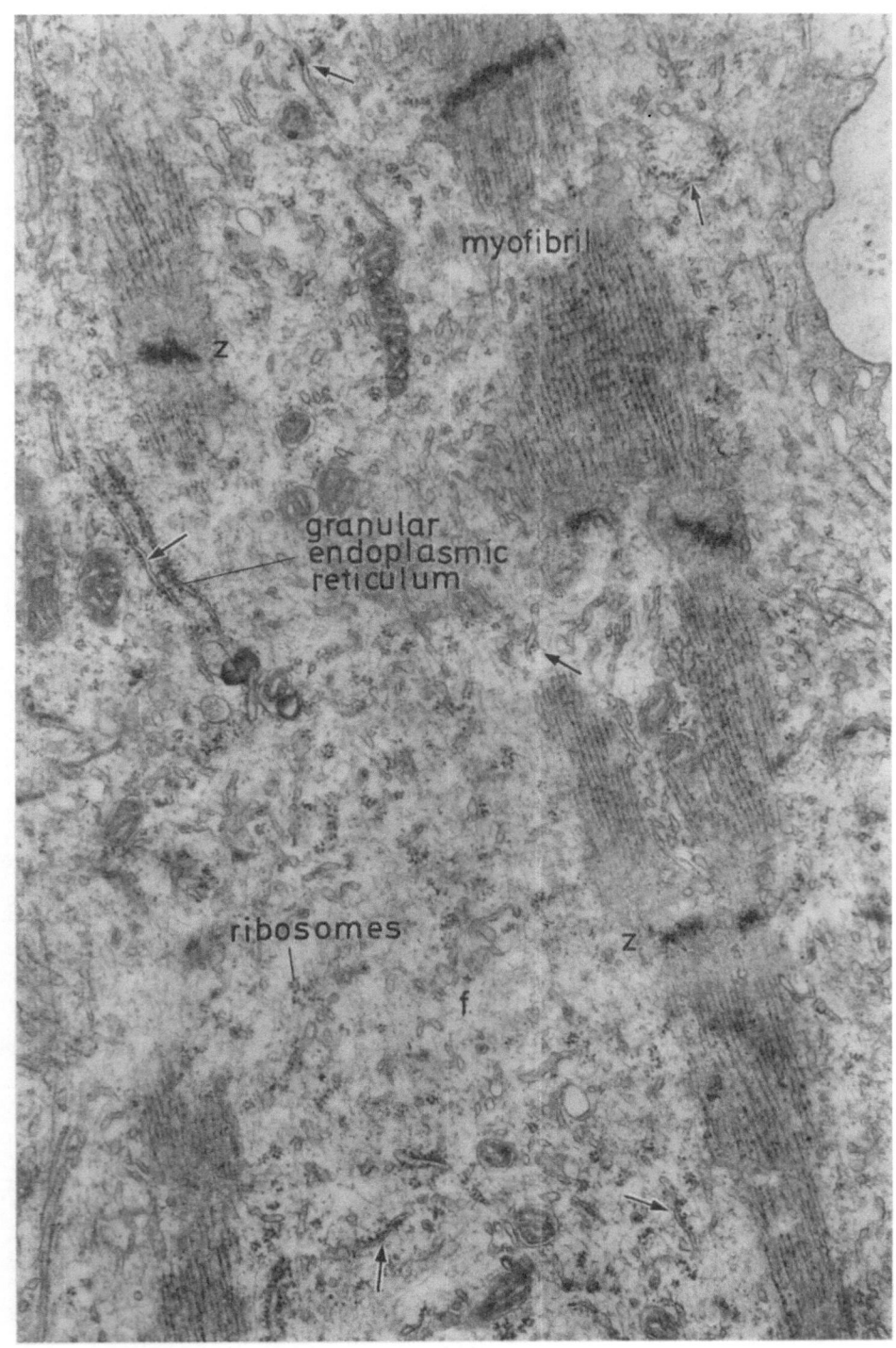

Fig. 3

Salamanders of the family *Ambystomidae* and certain anurans can regenerate a new lens from a fragment of the old lens. Members of the family *Salamandridae* are able to regrow a lens from the dorsal iris after complete removal of the old lens. The pupillary margin of the iris swells and a cavity appears that corresponds to the original cavity of the optic cup. Then, the iris cells depigment, nuclei and nucleoli enlarge and the growing mass takes the shape of a vesicle (REYER 1962). The cytoplasm of the cells becomes basophilic as new RNA is synthesized and the cells replicate their DNA and divide. Whether or not the dedifferentiated iris cells redifferentiate into lens fibers depends on their position. Some of the dividing cells remain in the iris epithelium without depigmenting completely, others replicate a number of times before they begin to synthesize alpha, beta and gamma crystallins, while others divide only once before forming lens cells (YAMADA and TAKATA 1963; TAKATA and YAMADA 1964; YAMADA 1966).

Dedifferentiation has been described during regeneration of some mammalian organs (see HAY 1966b). Regeneration of skeletal muscle is less dramatic in the mammal than in the amphibian, but the multinucleate fibers do seem to have the ability to produce mononucleate cells that proliferate to repair small gaps (BINTLIFF and WALKER, 1960; ALLBROOK, 1962) The onset of cell proliferation in mammalian liver after partial resection is remarkably rapid. Not very much cell dedifferentiation occurs in this case (BUCHER 1967). There is dispersion of cytoplasmic basophilic bodies on the first day, but the granular endoplasmic reticulum rapidly reforms. The proliferating cells synthesizing DNA and RNA have large nuclei and prominent nucleoli. Nevertheless, they are still recognizable as liver cells and seem to continue to make liver proteins. We can conclude, however, that when there is dedifferentiation in vertebrate regeneration, the process leads to and is part of the proliferative response of the tissue to injury or resection. Whether or not some dedifferentiation is necessary for resumption of proliferative activities by the cells of stable tissues will be the subject of the last section of this chapter.

III. Dedifferentiation in Invertebrate Regeneration

The widely held opinion that the differentiated state is irreversible by definition has nowhere influenced the thinking of the embryologist more than in the field of invertebrate regeneration. In many of the invertebrate tissues, there appear to be groups of basophilic cells which could conceivably serve as reserve cells in regeneration. For example, epidermis of most coelenterates contains numerous basophilic cells called interstitial cells. In an electron microscopic study, SLAUTTERBACK and FAWCETT (1959) reported that the interstitial cells contain numerous free ribosomes and little or no endoplasmic reticulum. They saw no reason to believe, however, that these cells did anything more than replace spent cnidoblasts. Such evidence as there is for participation of interstitial cells in Hydrozoan regeneration stems from light micro-

Fig. 3. Higher magnification electron micrograph of a portion of the cytoplasm of the dedifferentiating muscle fiber shown at B in Fig. 2. Myofibrils break apart in the A and I band regions rather than at the Z lines. The Z band agglutinates and disappears (Z). For a time, filamentous material seemingly derived from the myofilaments persists in the background cytoplasm (f). Cells such as these are not dying. The cytoplasm is acquiring free ribosomes and segments of granular endoplasmic reticulum. Configurations suggesting conversion of smooth-surfaced sarcoplasmic reticulum to granular reticulum can be seen (arrows). 30000 ×

scopic studies. After transection of a polyp, interstitial cells adjacent to the wound area seem to enlarge, divide and migrate from epidermis to endodermis. Cell counts suggest that the number of interstitial cells decreases in the rest of the body as the cells migrate to the wound area (TARDENT 1963). Distal movement of cells along the stalk can be visualized and photographed in the living material (STEINBERG 1955). Some of these cells are said to be interstitial cells and some are endodermal cells (TARDENT 1963). If any of the endodermal or epidermal cells transformed into basophilic cell types at this time, they would of course be counted as interstitial cells at the light microscope level. It is said that the interstitial cells are particularly sensitive to x-ray and that this is the reason why irradiation has a deleterious effect on regeneration in coelenterates. It would be just as reasonable to think that x-ray would prevent the proliferation of all cell types, however, not just that of interstitial cells. X-ray is known to inhibit mitosis in vertebrates, probably by blocking the transition from G-1 to S or G-2 to M (PUCK and STEFFEN 1963).

HAYNES and BURNETT (1963) have reopened the issue of the necessity of interstitial cells for regeneration in Hydra. During regeneration of polyps from fragments of pure endodermis lacking interstitial cells, glandular digestive cells dedifferentiate and give rise to basophilic cell types which proliferate and redifferentiate into cnidoblasts. Absorptive intestinal cells seem to give rise to epitheliomuscular cells without proliferating, but they lose the symbiont algae and other characteristics of digestive cells along the way (see BURNETT, this volume). These studies do not prove that dedifferentiation is part and parcel of normal regeneration in coelenterates possessing interstitial cells, but they call attention to the need of a reinvestigation of the subject using newer cytological methods and autoradiography to document the possible participation of the formed tissues in the process.

Of all the invertebrates the worms, particularly the flat worms, are most renowned for their possession of so-called reserve cells. It is said that the Turbellarians abound in large basophilic neoblasts which are totipotent and migrate to the wound surface to form the regeneration blastema (LENDER 1962, WOLFF 1962; STEPHAN-DuBOIS 1965). The evidence that reserve cells are requisite for planarian regeneration can be challenged, however. The cells in question are identified at the light microscope level by their basophilic cytoplasm after methyl green-pyronin staining. They are located in the mesoderm between the epidermis and the endodermis. Cell counts suggest that in normal regeneration, neoblasts are depleted from the proximal area of the body at the time that basophilic cells are seen accumulating at a cut surface. As we saw was the case for coelenterates also, it is said that planarians cannot regenerate after total body irradiation because of damage to these so-called neoblasts. If one end of the animals is irradiated, basophilic cell types seem to migrate from unirradiated regions of the animal to support regeneration in the irradiated part. These studies could not rule out the possibility that differentiated cell types in the unirradiated area of the body gave rise to some of the migrating basophilic cells by dedifferentiation. The stain used only reveals basophilic cells and would not show such transformations if they occurred.

HYMAN (1951) originally took the view that reserve cells were probably not involved in planarian regeneration and she has reviewed the earlier cytological literature along these lines. In particular, the results of LANG (1912) who used a variety of staining techniques to study cell structure are worth re-examining (Fig. 4). He reports in agreement with others that the wound epithelium derives entirely from old epidermis. The circular and longitudinal muscle layers under the epidermis depigment and seem to dedifferentiate to contribute cells to the regeneration blastema which is forming at the surface. The mesoderm of the planarian contains numerous

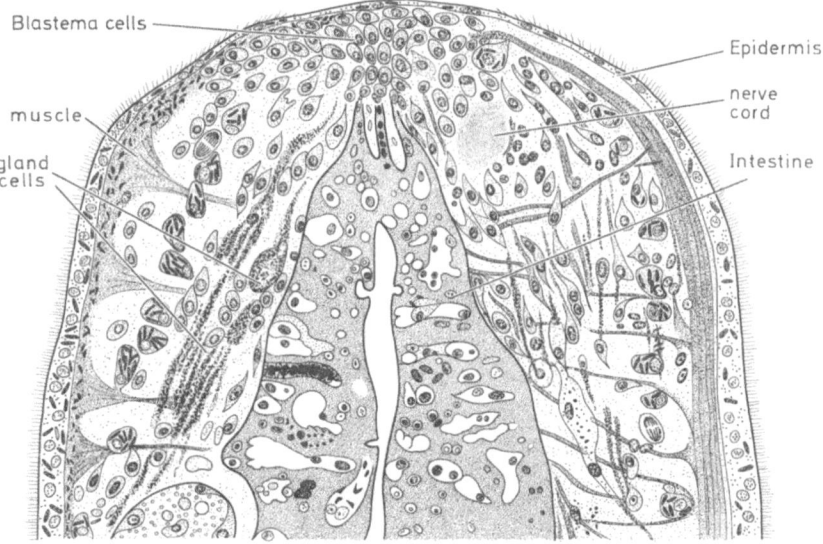

Fig. 4. Semidiagrammatic drawing of a longitudinal section through the anterior part of a regenerating planarian one day after amputation of the head. The formed tissues seem to give rise to the regeneration cells, much as in the case of the amphibian limb. (After LANG 1912)

gland cells which light microscopists have classified as predominantly serous and mucous in type. Under normal conditions they sent long necks up into the epidermis where their products are discharged to the exterior to give rise to the slime on the surface of the flatworm. These gland cells are depicted in various stages of dedifferentiation by LANG and there also seemed to be a contribution of cells from the gut and the kidney to the blastema.

We have recently confirmed LANG's results using the electron microscope to study regeneration in the species *Dugesia tigrina*. When the worms are transected proximal to the eyes, the tissues near the cut surface dissociate completely within a few hours.

Fig. 5. Low magnification electron micrograph of a lateral region proximal to the blastema in a regenerating planarian one day after amputation of the head. Some of the cells migrating toward the wound have already dedifferentiated, as at d. Others can still be recognized by the differentiated products they contain (gland cells) or by the cell shape (kidney cells). The tissue from *Dugesia* shown in Fig. 5—9 was fixed in glutaraldehyde followed by osmium tetroxide. Araldite sections were stained with uranyl acetate and lead citrate. 3300 ×

Fig. 6. Dedifferentiating muscle cell near the blastema in regenerating *Dugesia* one day after amputation of the head. This cell has a bilobed nucleus and abundant granular endoplasmic reticulum (arrows). The cell body would be basophilic as viewed in the light microscope, yet it is clearly part of a muscle cell. The myofibrillar portion is demarcated from the cell body by a cytoplasmic constriction. Z band-like material is recognizable (Z) as well as myofilaments. The myofibrillar portion may either degenerate or be cast off. Several nerve fibers appear at the upper right (n). 35000 ×

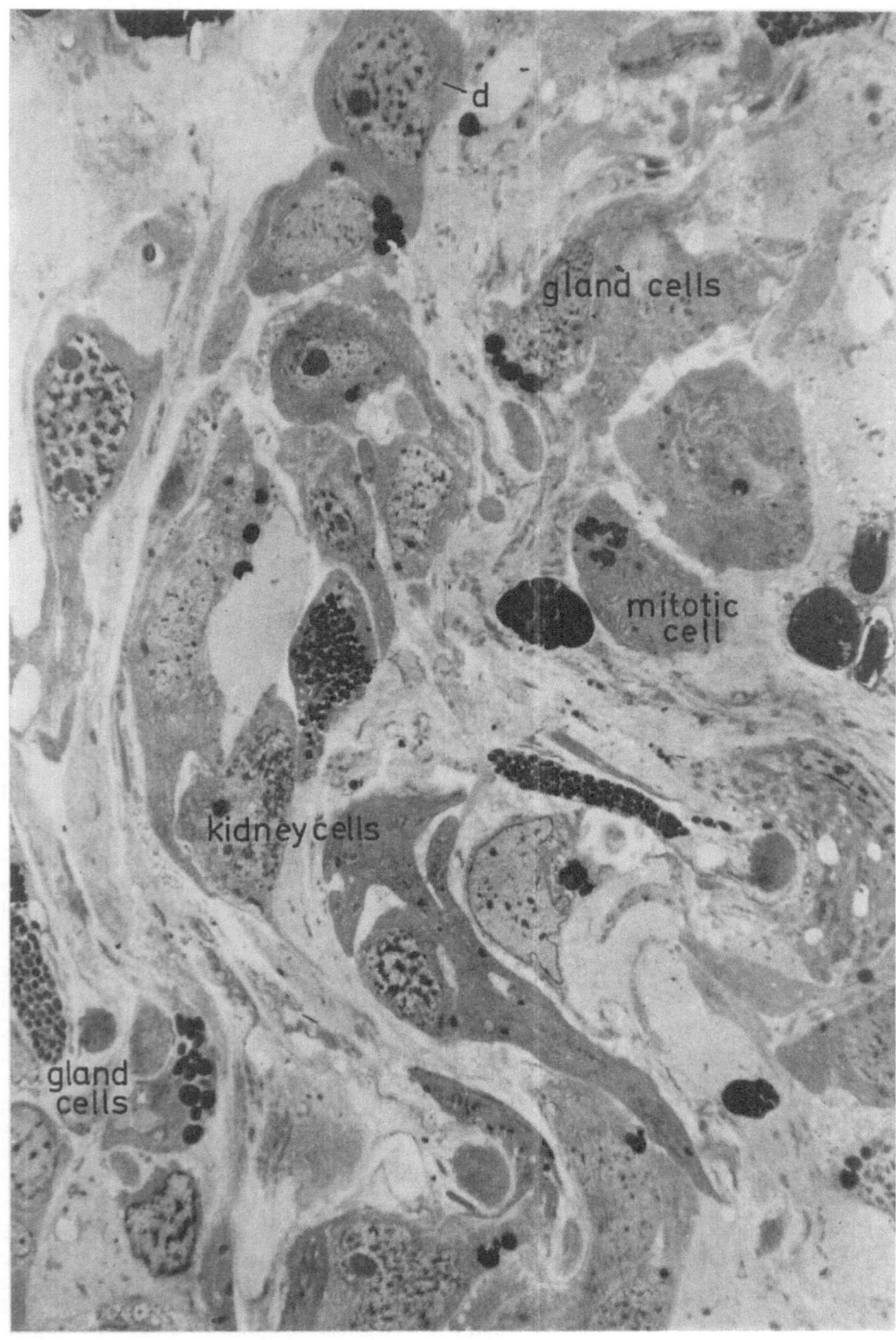

Fig. 5

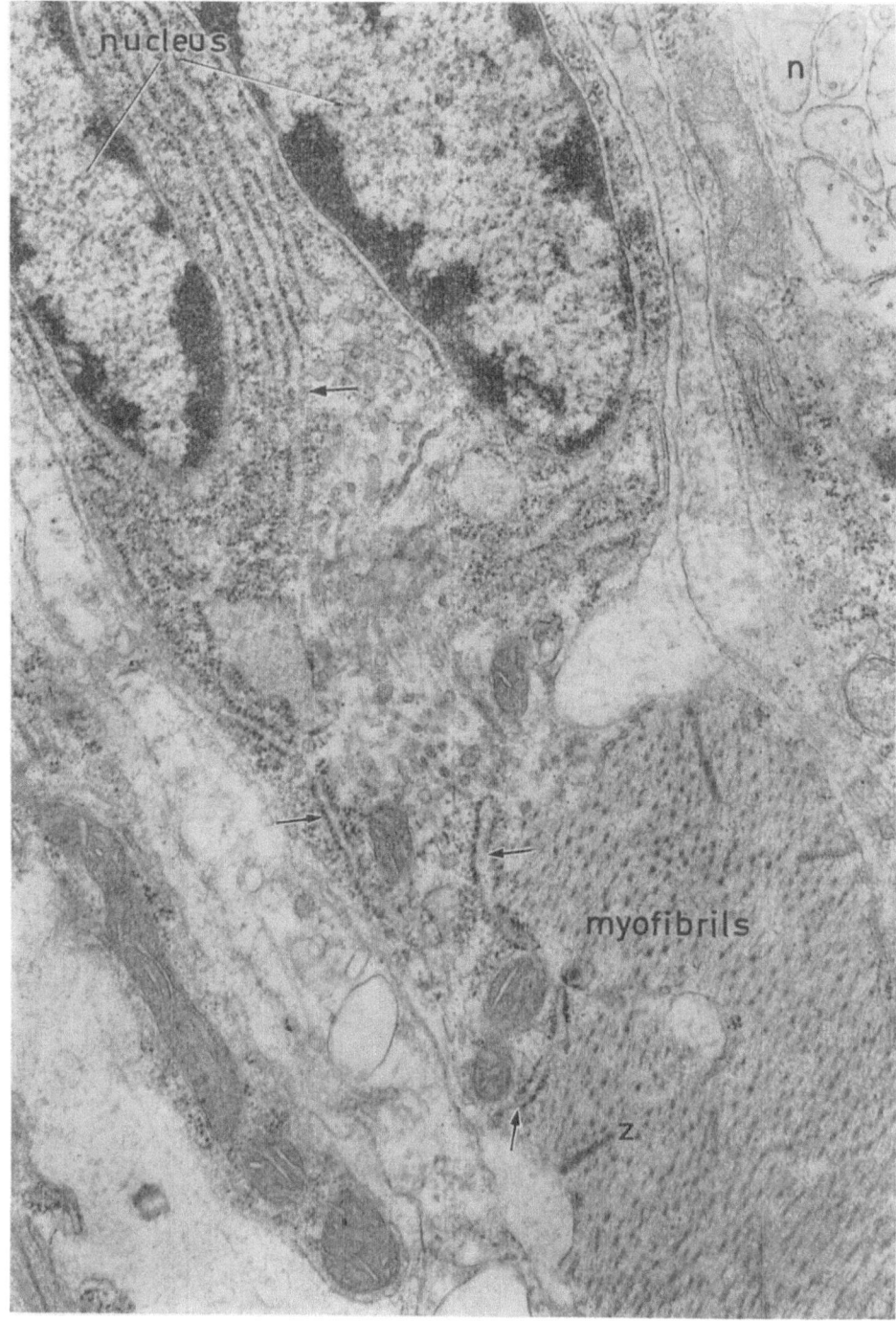

Fig. 6

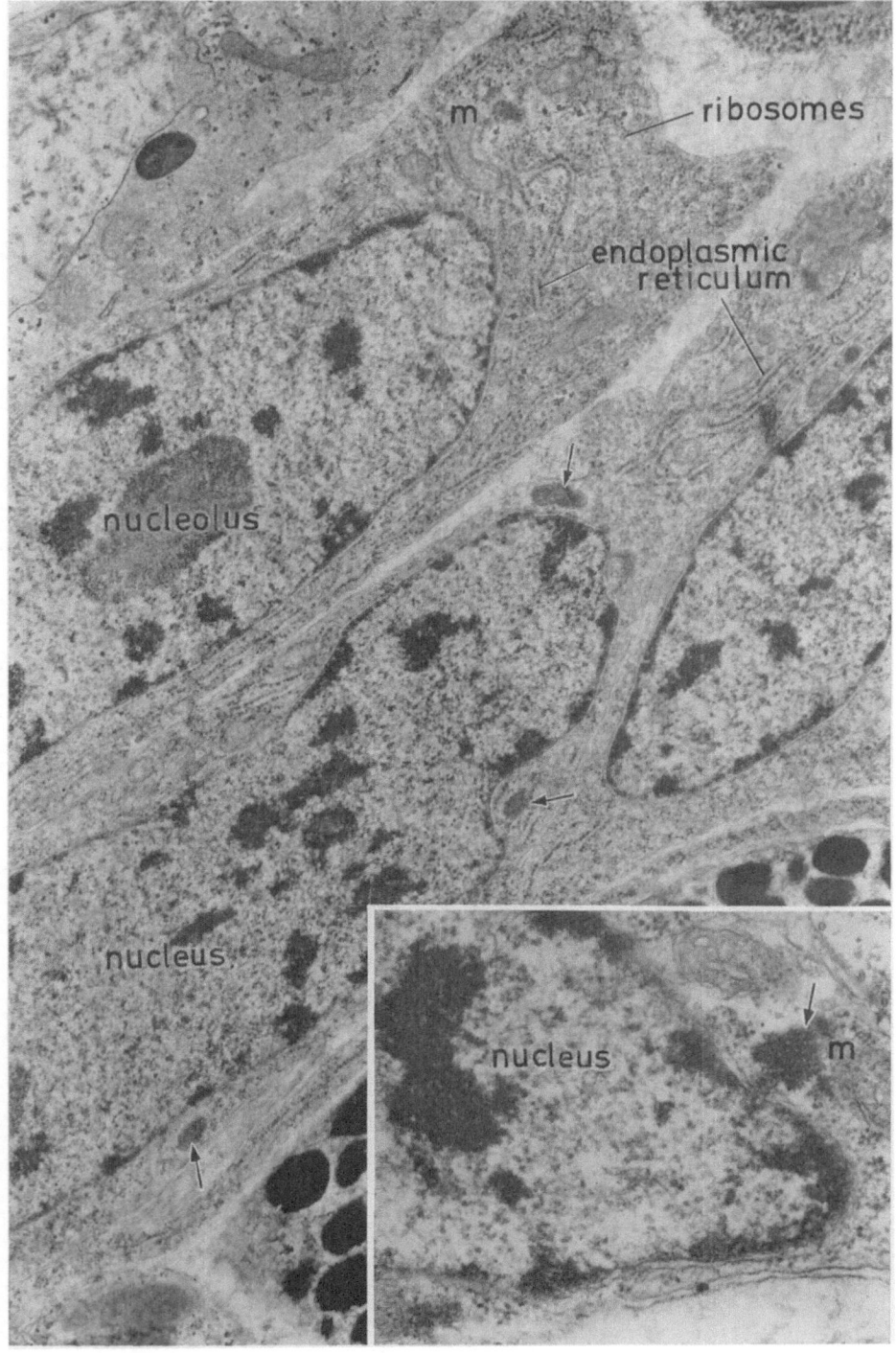

Fig. 7

Twenty-four hours after amputation, there is a gradient of tissue dedifferentiation extending proximally for a considerable distance and many cells are beginning to divide. In the electron micrographs, one is struck by the configuration of the cells. They do indeed seem to be migrating toward the tip as the earlier workers believed (Fig. 5). What seems clear from these observations, is that some of the cells which are involved in the migration are obviously still differentiated at the beginning. Gland cells, muscle cells and even kidney cells are recognizable in the migrating masses. The syncytial gastrointestinal tract breaks up at the tip and multinucleate and binucleate cells derive from it which then further dissociate into mononucleate cells much as does skeletal muscle in the amphibian limb. As in the case of the amphibian limb, it is possible to find all stages of morphological dedifferentiation in tissues proximal to the cut. The muscle cells lose myofibrils and the cytoplasm can be seen to contain large amounts of granular endoplasmic reticulum and free ribosomes (Fig. 6).

The amount of granular endoplasmic reticulum in the undifferentiated blastema cells at the tip of the regenerate varies (Fig. 7). The more distally located cells are the least differentiated in appearance. The cytoplasm contains mitochondria, numerous free ribosomes and few cytoplasmic membranes. A remarkable feature which is not observed in the amphibian blastema cell is the appearance of dense cytoplasmic bodies in association with the nuclear pores (arrows Fig. 7), reminiscent of the "nuclear extrusions" reported in oocytes (KESSEL 1966). Such extrusions do not occur in more differentiated cells and perhaps reflect the stepped up nucleocytoplasmic exchange that is occurring during the phase of rapid cell replication. The planarian does not grow very markedly in length during the first week after amputation and the blastema derives by reorganization of the internal components. Nevertheless, cell divisions are common (Fig. 8). The blastema is well formed by 24 hours and by 72 hours it has already started to redifferentiate. Growth of the worm to its previous size is accomplished gradually through repeated division of the cells in the immediate area of the blastema and in the proximal regions. The loss of cytoplasmic specializations, swelling of nuclei, enlargement of nucleoli and acquisition of other features associated with proliferative activity observed in cells near the blastema and at proximal levels in the amputated worm satisfy our criteria for dedifferentiation.

A second rather damaging line of evidence against the reserve cell theory comes from the fact that we were unable to find any large reserve of relatively undifferentiated cells in the adult planarian. The mature planarian which is not undergoing asexual reproduction possesses discrete gonads, excretory and digestive systems with which so-called neoblasts are not associated. Cells which would be counted intensely basophilic as viewed in a light microscope after methyl green-pyronin staining are primarily of gland cell morphology and occur in the mesoderm. These cells are differentiated in the sense that they possess an extensive granular endoplasmic reticulum, Golgi complex and accumulations of secretory products. Even as viewed in

---

Fig. 7. Blastema cells in *Dugesia* at the regeneration stage depicted in Fig. 4. These basophilic cells contain varying amounts of granular endoplasmic reticulum and considerable numbers of free ribosomes. Nucleoli are prominent. Dense chromatin-like bodies (arrows) are seen in the cytoplasm, often connected to pores in the nuclear envelope (inset). These chromatoid bodies may contribute stuff to the mitochondrial matrix (m). The blastema cells do not contain differentiated products, such as myofibrils or secretory granules. 20000 × ; inset 40 000 ×

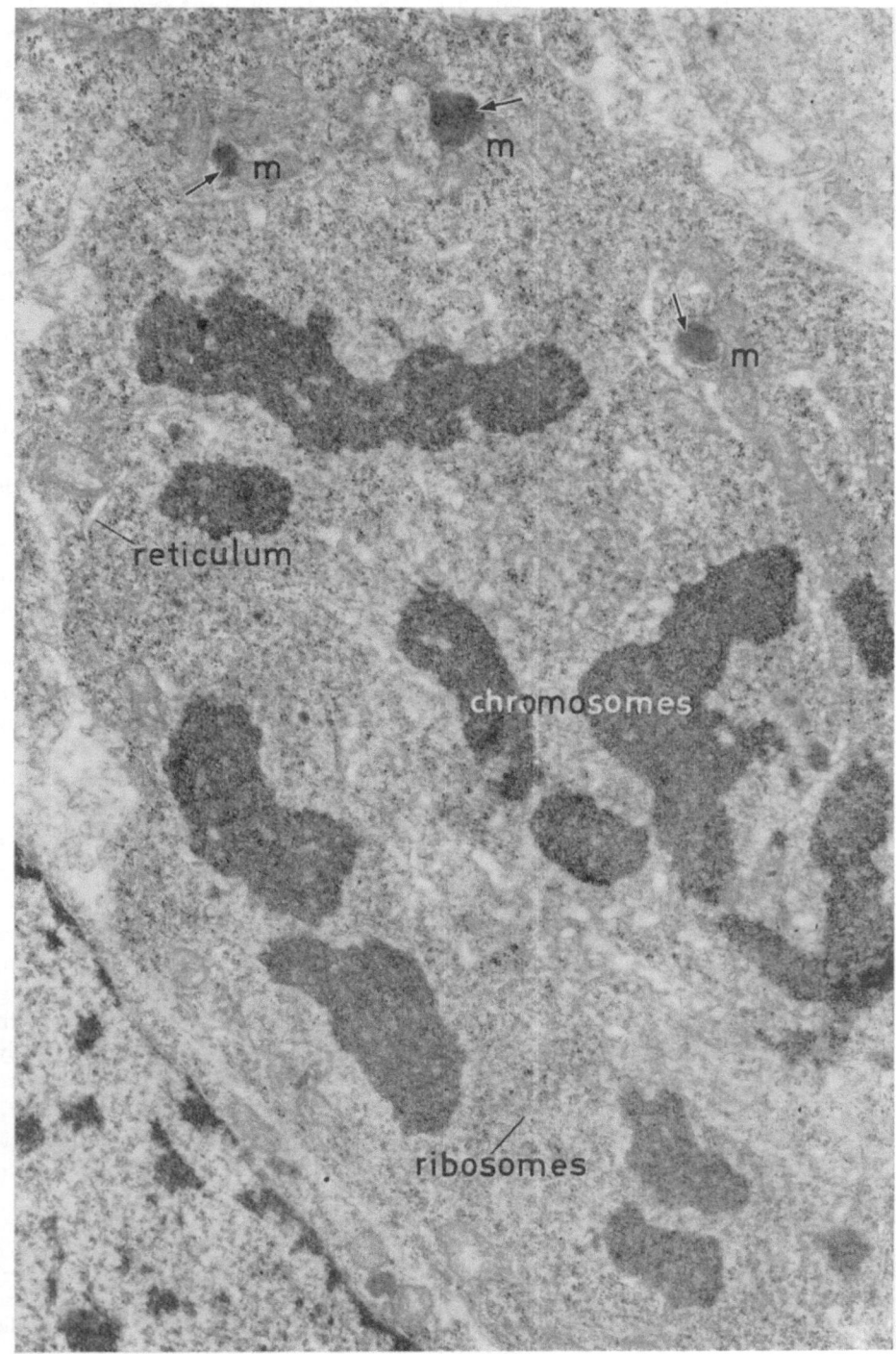

Fig. 8

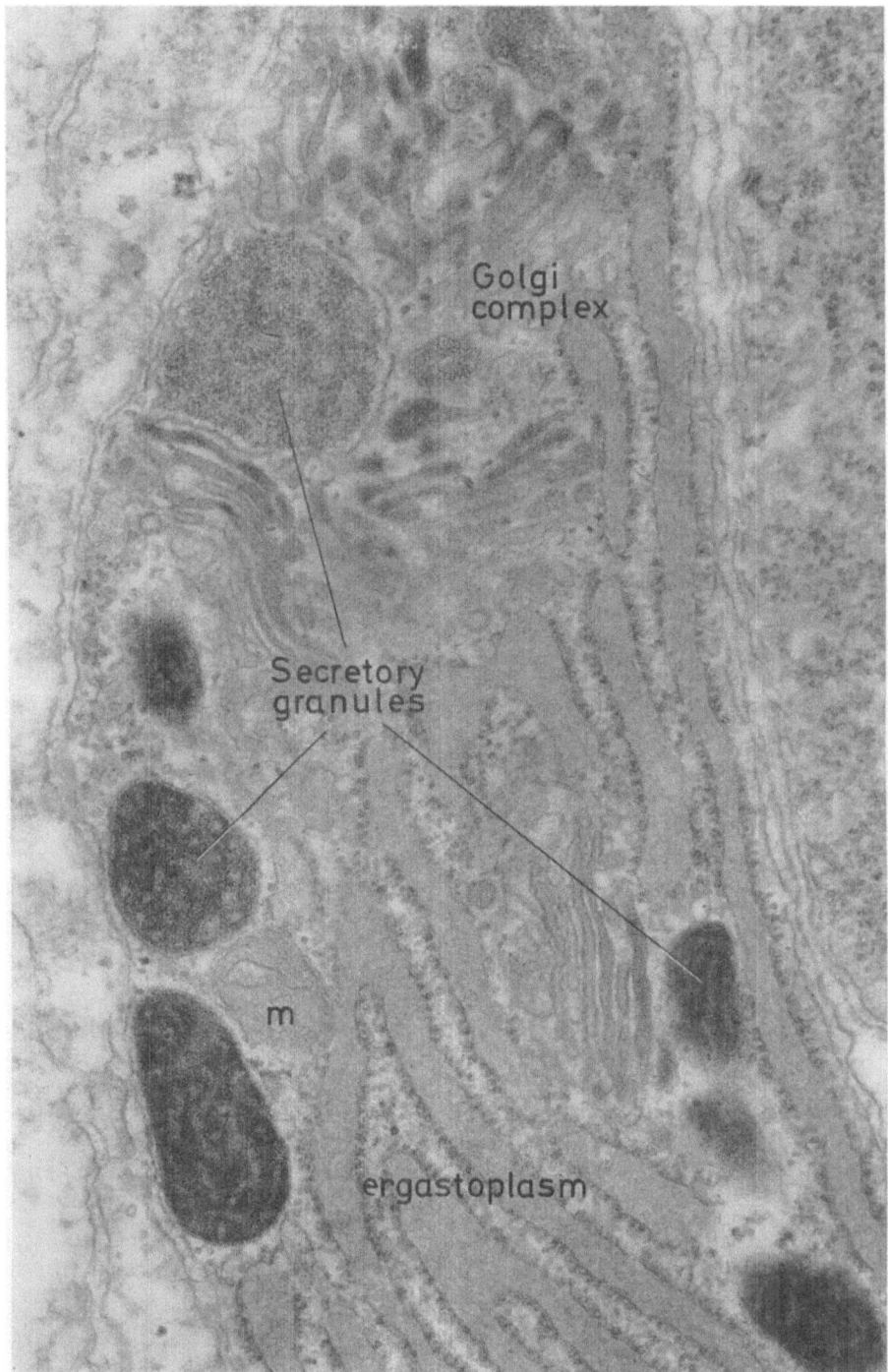

Fig. 9

the light microscope, the basophilic cell types depicted by LENDER (1962) and others have definite clear areas in the cytoplasm next to the nucleus, much as do the plasma cells of mammals. These clear areas correspond to the Golgi region of the cell and contain secretory granules (Fig. 9). It is difficult to avoid the conclusion that most, if not all the so-called reserve cells in the mesoderm of the adult planarian are merely stages in the secretory cycles of the serous and mucous cells which are continually producing the slimes and glues that characterize the surface of this animal.

IV. Does Metaplasia Occur in Regenerating Systems?

If we define cytodifferentiation in terms of the specializations of cells that equip them for a somatic cell function, then it is easy to see that in many regenerating systems cells are capable of reversing seemingly stable patterns of somatic differentiation. Myofibrils and extracellular matrices can be disposed of and cells can assume the morphology of growing embryonic cells. *There is no necessity to invoke the concept of "reserve cells" to explain regeneration.* The formed tissues can and do give rise to regeneration cells in many instances. But in the process, do the cells broaden the immediate spectrum of expression of their developmental potentialities? This is an old question and in rephrasing it, we now inquire as to whether or not genetic control mechanisms remain stable under environmental conditions which lead to the loss of specialized cell products. During the period of proliferation, when differentiated products decrease or disappear, does the messenger RNA associated with the maintenance or manufacture of such products continue to be synthesized? Alternatively, one might ask whether or not regulator genes are propagated, in the absence of the synthesis of specialized cell products, in a form which nevertheless will bias a cell to redifferentiate along the old pattern. It is tempting to think that the simplified blastema cells which appear in regeneration are at least pluripotent within the restrictions of the particular environment (SCHOTTÉ 1940). Nevertheless, the direct answer to this question rests almost entirely on our ability to demonstrate the presence or absence of metaplasia in regenerating systems, that is, to show whether or not cells of one line, such as muscle, can give rise to a different line, such as cartilage. This is a difficult task and the answers that are available are not necessarily fair ones.

Fig. 8. Mitosis in a basophilic cell near the blastema in *Dugesia* one day after amputation of the head. This cell could have derived from any of the formed cell types of the amputated worm. It is relatively undifferentiated in appearance and contains numerous free ribosomes and some granular endoplasmic reticulum. Chromatoid bodies (arrows Fig. 7) can be discerned in the cytoplasm (arrows) often associated with mitochondria (m). 20000 ×

Fig. 9. Portion of the cytoplasm of one of the gland cells in the "mesoderm" of the adult planarian. Some gland cells are engorged with secretory products and are easily identified at the light microscope level. Others, such as this one, contain only a few secretory granules and extensive endoplasmic reticulum (ergastoplasm). They would be recognized as intensely basophilic as viewed in the light microscope and probably classified as "reserve cells". The Golgi complex is prominent, however, and can be seen as a negative image in the light microscope, thus revealing the true identity of the cells. In the Golgi zone, secretory products can be seen in various stages of "packaging" for release to the exterior. The products reach the surface of the worm by means of long cytoplasmic processes that extend into the epidermis (not shown). A mitochondrion is labelled (m). Chromatoid bodies, which are characteristic of growing cells (Figs. 7, 8), are not present in these differentiated gland cells. 56000 ×

It has proved impossible to demonstrate outright metaplasia in the regenerating amphibian limb under "normal" conditions. In the newt, redifferentiation of cartilage next to the cut end of the amputated humerus begins 15 to 20 days after amputation at the time that the newly formed blastema has just begun its actual outgrowth. What this means is that at the time the regenerate is in the cone stage, redifferentiation has in fact begun and the phase of blastema formation as such is over. Experiments tracing the fate of cartilage transplants in the limb have been interpreted as proof that redifferentiation is true to original cell type (see STEEN 1967), but it must be remembered that the environments of the early cone, paddle and fingerbud stages of limb regeneration all favor cartilage differentiation. The entire skeleton is formed before any muscle appears that can be recognized in the light microscope. It has not proved possible to follow tritium labeled or polyploid transplants this long. The question would be easier to answer if muscle transplants were employed, then, to see if cells derived from muscle can redifferentiate into cartilage. The problem has proved difficult to attack experimentally because muscle is a mixed tissue as contrasted to cartilage, which can be dissected clean of other cell types. Supernumerary limb formation under certain experimental conditions affords indirect evidence that cell transformations are possible in the regenerating limbs (TRAMPUSCH 1965). There is quite good evidence, moreover, that it is not necessary to have cells derived from bone or cartilage in the limb for the skeleton to reform. Earlier investigators (WEISS 1939; THORNTON 1942) removed the old skeleton in its entirety. The limb regenerates after amputation and in the new distal portion, a new cartilaginous skeleton forms which is not in continuity with the trunk. A large population of dedifferentiated cells derives from the muscle and contributes to the blastema (THORNTON 1942), but it is possible that the skeleton arises only from the muscle fibroblasts. While this in itself would certainly have to be considered a cell transformation, metaplasia of fibroblasts to cartilage cells is common enough in vertebrates so as not to be accorded much general significance.*

---

* *Note Added in Proof.* In the interim since this chapter went to press, a paper by OBERPRILLER (1967) has appeared which provides another example of a transformation of this general type in a regenerating system and gives further evidence that the young blastema offers a cellular environment conducive to chondrogenesis. Blastemata from regenerating newt intestines were labeled with tritiated thymidine and transplanted to regenerating limbs. Heavily labeled chondroblasts, which could have come either from dedifferentiated smooth muscle cells or from intestinal fibroblasts, were detected by autoradiography 6—10 days after graft implantation. It is difficult to evaluate the significance of the light labeling (several silver grains per nucleus) observed over blastema cells in this experiment and in the experiments of STEEN (1967) and ROSE and ROSE (1965). In the latter case, large numbers of lightly labeled blastema cells were described in limbs that had received transplants of labeled epidermis, yet there is considerable evidence against such an epidermal origin of the blastema (see CHALKLEY 1954; HAY and FISCHMAN 1961). STEEN (1967) counted as originating from grafts, cells that not only were heavily labeled by tritiated thymidine but also were marked by polyploidy. He observed and chose to ignore light labeling of what were apparently host (non-polyploid) cells following the transplantation of radioactive polyploid tissue. Neither simultaneous administration of hot and cold thymidine, nor study of controls receiving dead grafts (ROSE and ROSE 1965; OBERPRILLER 1967), would rule out the possibility that light labeling observed in long term experiments is the result of reutilization of tritiated by-products given off by living grafts long after the cold thymidine or dead graft would have disappeared.

There is striking evidence of a dramatic transformation from one cell type to another in the case of lens regeneration in the salamander. In *Triturus*, as we noted earlier the dorsal iris can give rise to a new lens, yet the iris derived originally from the brain and has little in common with the ectoderm epidermis that gave rise to the lens in the embryo. TAKATA, ALBRIGHT, and YAMADA (1964) have taken the view that the replication of genetic material which occurs during proliferation of the iris epithelium is a necessary part of the transformation of iris cells to lens cells. This system, however, is far more complex than it appears on the surface. Only members of the family *Salamandridae* have the ability to regenerate lens from iris and only the dorsal iris has the capacity to give rise to a new lens (REYER 1962).

The experiments that seem to prove the capacity of coelenterate cells for total metaplasia during regeneration (HAYNES and BURNETT 1963) have been done under conditions almost as artifactual as those involved in transplanting a somatic cell nucleus to an unfertilized egg (GURDON 1964) or placing a carrot cell in coconut milk (STEWARD 1963). The fact that *Hydra viridis* can regenerate a whole animal from an endodermal fragment is remarkable, nevertheless. Moreover, this system may provide the only example of a significant cell transformation in an animal in the absence of cell division (see BURNETT, this volume). The only possible example in vertebrates reported to date is the transformation of muscle and other cell types to macrophages said to occur *in vitro* (see WEISS 1950).

The regeneration blastema of planarians seems to have unlimited developmental capacity (see WOLFF 1962). We do not know any more about the developmental potencies of the individual blastema cells, however, than about those of the cells that arise during amphibian limb regeneration. If, as seems likely, blastema cells derive from all the tissues of the stump then it is conceivable that they redifferentiate according to tissue of origin as BANDIER (1937) believed. There have been claims of epithelial metaplasia in regeneration of arthropod appendages, but these are open to doubt (NEEDHAM 1965).

It seems appropriate to conclude this discussion with some reference to mammalian regeneration, where absence of metaplasia seems to be the rule even though dedifferentiation is not uncommon. Regeneration of glands such as the liver or pancreas involves replacement of the same cell type by pre-existing cells without much dedifferentiation. Skeletal muscle fibers, however, do seem to dedifferentiate in response to injury and the basophilic mononucleate cells derived thereby proliferate, meet with other mononucleate cells, fuse and repair the gap if it is not too large. Cells derived from myocytes do not differentiate into other cell types in the mammal, nor do fibroblasts seem able to give rise to muscle cells, even though, as we have noted, fibroblasts probably can form bone or cartilage in all vertebrates. Thus, we can conclude that dedifferentiation may be associated with transformation of one cell type to another, as in amphibian lens regeneration and possibly in the case of the limb, but in some cases, especially in mammalian regeneration, the proliferating cells probably redifferentiate according to their tissue of origin. For most regenerating systems, the definite answer to the question raised in this section remains to be obtained. Nevertheless, the fact remains that the previously differentiated cells which compose the blastema have taken up new relations with one another and do produce new structures of which they have never before been a part, such as the eye of a regenerating worm or finger of a regenerating limb.

V. Is Dedifferentiation Necessary for Cell Division?

What we can say with certainty is that dedifferentiation is almost invariably accompanied by cell proliferation in regenerating systems. It is, in a sense, the cell's response to a strong stimulus to grow. In the amphibian limb, as we have seen, skeletal muscle breaks up into mononucleate cells and, in planarians, the syncytial gut epithelium dissociates. It might be rightly argued that in both these cases, the syncytial state was incompatible with cell proliferation. Similarly, in other tissue types, such as cartilage, differentiated products might interfere with cell division. In fact, if we are to believe some of the generalizations which have appeared in recent years regarding the incompatibility of cell differentiation and cell proliferation (see HOLTZER 1963), then it might seem reasonable to conclude that dedifferentiation is absolutely necessary for proliferation of the cells in regenerating systems. We have, however, already seen that differentiated gland cells can divide and if we turn now to the phenomenon of physiological regeneration, it will become obvious that many types of differentiated cells proliferate in the growing organism.

The cells of the adult vertebrate can be classified into three groups according to the degree to which they are renewed under physiological conditions. (1) *A labile group* which consists of blood cells and the surface epithelia. There is a well-defined progenitor cycle, but the cells in the growth zone are partially differentiated. In the case of the bone marrow, for example, the cells which incorporate tritiated thymidine under normal conditions are the proerythroblasts, erythroblasts, promyelocytes and myelocytes. They comprise about 25% of the bone marrow population and are easily recognized because they have begun to make cell-specific proteins. Autoradiographic studies prove that these partially differentiated cells, not the reticular cells, normally perpetuate the marrow population (see HAY 1966b). The cells of the intestinal crypts must also be regarded as partially differentiated and there can be no doubt that the dividing cells in the *stratum germinativum* of the skin have synthesized cell-specific tonofilaments that are now believed to be composed of keratin. (2) *A stable cell group* which has been characterized by LEBLOND, MESSIER, and KOPRIWA (1959) as expanding because some tritiated thymidine incorporation may occur in these tissues even in relatively mature adults. The epithelial gland cells, such as occur in liver, kidney and pancreas, and the connective tissue secretory cells fall into this group. The tissue is usually capable of interstitial growth and the end-product cell is not necessarily terminal. Regardless of the arguments as to the synthesis of cell-specific proteins by proliferating cells in culture (CAHN and LASHER 1967; see also CAHN, WHITTAKER, this volume), there seems little doubt that in the growing animal, cartilage cells do continue to divide, as do liver cells, kidney cells, fibroblasts and pancreatic acinar cells.

(3) *A static group* is said to show no cell division in the normal adult. These are the muscle and nerve cells. While it seems quite clear that differentiated neurons never divide, the nuclei do probably retain the ability to make DNA (GURDON 1967). We have already seen that skeletal muscle fibers can dissociate into individual cells which are capable of proliferating, even in the mammal. It is sometimes argued that synthesis of cell-specific proteins, such as myofibrils, turns off DNA synthesis in the developing skeletal muscle fiber. It seems far more likely, however, that it is the syncytial state of the fiber which is incompatible with further nuclear growth. Cardiac muscle cells

do not form syncytia and they continue to divide even while developing quite respectable myofibrils as determined by electron microscopy (MANASEK 1967). The possibility that cardiac muscle cells have some ability to divide following injury in the adult has not been adequately investigated.

When one considers the spectrum of physiological regeneration in the adult and the overall development of the embryo, it seems impossible to argue that synthesis of cell-specific proteins is in itself incompatible with cell division. There is obviously a whole set of factors antagonistic to growth in the mature organism in the static and stable tissues referred to above, but this may have only to do with the cytoarchitecture of these particular tissues and may be mediated by control mechanisms related only indirectly to the so-called cell-specific proteins which arise during particular phases of development. While it is not unreasonable to conclude that DNA cannot simultaneously replicate itself and produce RNA, the messenger RNA of most multicellular organisms is stable enough to survive the 6—8 hour period of DNA replication required by most cells to divide. This 6—8 hour "S" period is remarkably constant in animals. The only good example of more rapid DNA synthesis is seen in a highly specialized cell system, the cleaving oocyte (see GURDON 1967). One might argue, however, that the blastema cell must spend the whole of its "G_1" interphase making proteins needed for cell replication and that formation of the blastema reflects the high rate of cell division in reparative regeneration.

It is perhaps just as wrong to generalize from one regenerating system to another as it is to generalize about cell proliferation from studies *in vitro* of skeletal muscle and cartilage cell renewal. We have in our studies of regeneration tended to ignore those systems in which formation of a blastema is not essential. Thus, in assessing the necessity of dedifferentiation in reparative regeneration, we must keep in mind the fact that 30% of the remaining cells are in the process of replicating their DNA 20 hours after excision of two-thirds of the rat liver. And they do this without significant cell dedifferentiation. In a regenerating newt limb, not many more than a third of the cells are synthesizing DNA at any one time. The figure is higher for larval *Ambystoma*, but it is not that much higher. The growth process is, however, sustained for a considerably longer period in regeneration of an appendage as compared with that of a gland. One might argue that in an appendage composed of diverse tissues, formation of a blastema is necessary to provide enough cells for reinstatement of a complex morphological pattern. Perhaps the same is true for regeneration of parts of a worm. Coelenterates are said not to form a blastema during regeneration after a cut, but asexual reproduction does involve production of a bud of embryonic-appearing cells in these organisms. In limb regeneration, the mass or size of the blastema seems to be one of the critical factors in determining the success of the regrowth. It may be fair to conclude, then, that initiation of sustained DNA synthesis either requires, or brings about inadvertently, a loss of those special cell features which we associate with the differentiated state.

Returning to the theme of this volume, whether or not it is necessary in a theory of differentiation to assume that the differentiated state is irreversible, it is clear from consideration of the regenerating systems that a number of cell types readily revert to more embryonic-appearing cells under conditions of trauma and start to divide. The epithelia and epithelial glands retain recognizable vestiges of their former state of differentiation, but the tissues derived from mesenchyme give rise to simplified cell types that are indistinguishable from each other, at least in the case of the amphibian limb. In viewing the whole spectrum regenerative processes in animals, it is obvious that the so-called "stability" of the differentiated state is dispensable or modifiable in response to an adequate growth stimulus, but at the same time it is possible to imagine the existence of genetic control mechanisms which regulate the behavior of the cells under even the most extreme conditions of dedifferentiation and lead either to controlled metaplasia or to reinstatement of the original pattern of differentiation. While, as we indicated in the introduction, there is ample evidence that the cytoplasm of the ovum can cause a differentiated nucleus to resume the metabolic activities which characterize the developing embryo and lead it in some cases to re-express its "totipotency", much remains to be learned about the levels at which the cell normally expresses its inherent developmental capacity. Dedifferentiation and cell proliferation lead to the production of a whole animal from somatic cells during regeneration in certain of the invertebrates, but in the vertebrates, total reproduction of this kind always involves the gametes. Thus, although we may admit that all cell nuclei are possibly "totipotent", experiments on regeneration in the vertebrate seem to attest to the conservatism with which such potentials are expressed under normal conditions, even in the face of massive cytoplasmic dedifferentiation and DNA replication. The fact that muscle cells do not give rise to an embryo in the interior of a regenerating limb should not lead us, however, to shrug off the phenomenon of dedifferentiation as merely "morphological" or to regard it as some kind of a deception. The reversibility of differentiation exhibited by regeneration cells must be accounted for in any scheme that purports to explain cytodifferentiation.

VI. Summary

When vertebrate appendages composed of mixed cell types regenerate, a blastema of relatively undifferentiated-appearing cells usually forms. The blastema cells arise by dedifferentiation of the formed tissues. Electron microscopic observations confirm the impression that differentiated cell products are lost. Cell organelles are simplified, as in proliferating embryonic cells, and numerous free ribosomes usually appear in the cytoplasm. Thus, dedifferentiation in regenerating systems can be defined as consisting of (1) partial or complete loss of products associated with the differentiated state, such as myofibrils and extracellular secretions, concomitant with (2) acquisition of nuclear and cytoplasmic characteristics that are compatible with cell division, such as renewed RNA and DNA synthesis, enlargement of nuclei, and increased cytoplasmic basophilia. The same general changes can be noted in regeneration of the lens and intestine of the amphibian, but higher vertebrates seem limited in their ability to repair organs as complex as these.

It is generally assumed that the lower invertebrates capable of asexual reproduction and whole body regeneration, owe these remarkable capacities to their possession of so-called reserve cells. In coelenterates, the interstitial cells that normally give rise to cnidoblasts have been equated to reserve cells because they possess numerous ribosomes and are undifferentiated in their fine structure. It can be argued, however, that these basophilic cells are not necessary for regeneration (see BURNETT, this volume). Of all the invertebrates, the worms, particularly the flatworms, are most renowned for their possession of so-called reserve cells. Evidence is presented here which suggests that the large basophilic cells considered to be neoblasts in the mature planarian are, in fact, active gland cells. Studies with the electron microscope also suggest that during regeneration from an anterior cut, the worm forms a blastema of basophilic cells by dedifferentiation of the formed tissues of the body. Some of the blastema cells are completely undifferentiated in appearance and have cytoplasmic extrusions associated with nuclear pores. Others have abundant endoplasmic reticulum. Studies of DNA synthesis have not yet been done, but it is clear that here, as in the vertebrates, the process of dedifferentiation is accompanied by mitosis.

Some general questions are raised and discussed briefly. The widespread occurrence of dedifferentiation in regeneration is emphasized, but it is also pointed out that epithelial organs such as the liver may regenerate without a very striking phase of dedifferentiation. The question as to whether or not metaplasia occurs in regeneration is raised and left with a very incomplete answer. Lens regeneration provides the most striking example of transformation of one cell type to another in the amphibian. We also ask whether or not dedifferentiation of stable cell types is required for resumption of DNA synthesis and we come to the conclusion that cell differentiation and cell division are not necessarily incompatible. Differentiated cartilage cells, fibroblasts, kidney cells, smooth and cardiac muscle cells and many gland cells can and do divide in the embryo and during physiological renewal of tissues in the adult. What may be special about the regenerating systems that form blastemata by "massive" cell dedifferentiation is that the architecture of the removed part was particularly complex. In limb regeneration, the mass or size of the blastema seems to be one of the critical factors in determining the success of the regrowth. Even though the subsequent behavior of the blastema cells seems conservative in terms of re-expression of "total developmental capacity", the phenomenon of dedifferentiation is remarkable and must be accounted for in any theory of cytodifferentiation.

References

Allbrook, D.: An electron microscopic study of regenerating skeletal muscle. J. Anat. 96, 137—152 (1962).

Bandier, J.: Histologische Untersuchungen über die Regeneration von Landplanarien. Arch. Entwickl.-Mech. Org. 135, 316—348 (1937).

Bintliff, S., and B. E. Walker: Radioautographic study of skeletal muscle regeneration. Amer. J. Anat. 106, 233—245 (1960).

Bucher, L. R.: Experimental aspects of hepatic regeneration. New Engl. J. Med. 13, 686—696 (1967).

Butler, E. G.: The effects of X-radiation on the regeneration of the forelimbs of Amblystoma larvae. J. exp. Zool. 65, 271—315 (1933).

Cahn, R. D., and R. Lasher: Simultaneous synthesis of DNA and specialized cellular products by differentiating cartilage cells in vitro. Proc. nat. Acad. Sic. (Wash.) 58, 1131—1138 (1967).

Chalkley, D. T.: A quantitative histological analysis of forelimb regeneration in Triturus viridescens. J. Morph. 94, 21—70 (1954).

Deck, J. D.: Cytological changes in newt limb muscle after limb amputation. Anat. Rec. 160, 338 (1968).

Fraisse, P.: Die Regeneration von Geweben und Organen bei Wirbeltieren. Cassel-Berlin: T. Fischer 1885.

Gurdon, J. B.: The transplantation of living cell nulcei. In: Advances in Morphogenesis, (Ed.: M. Abercrombie and J. Brachet) Vol. 4, pp. 1—43. New York: Academie Press 1964.

GURDON, J. B.: On the origin and persistence of a cytoplasmic state inducing nuclear DNA synthesis in frog's eggs. Proc. nat. Acad. Sci. (Wash.) 58, 545—552 (1967).

HAY, E. D.: Cytological studies of dedifferentiation and differentiation in regenerating amphibian limbs. In: Regeneration (Ed.: D. RUDNICK), pp. 177—210. New York: The Ronald Press Co. 1962.

— Embryological origin of the tissues. In: Textbook of histology. (Ed.: R. O. GREEP), pp. 56—73. New York: McGraw Hill Book Co. 1966a.

— Regeneration. Biology study series. New York: Holt, Rinehart and Winston, Inc. 1966b.

—, and D. A. FISCHMAN: Origin of the blastema in regenerating limbs of the newt, *Triturus viridescens*. An autoradiographic study using tritiated thymidine to follow cell proliferation and migration. Develop. Biol. 3, 26—59 (1961).

HAYNES, J., and A. L. BURNETT: Dedifferentiation and redifferentiation of cells in *Hydra viridis*. Science 142, 1481—1483 (1963).

HOLTZER, H.: Mitosis and cell transformations. In: General physiology of cell Specialization (Ed.: D. MAZIA and A. TYLER). New York: McGraw-Hill Co. 1963.

HYMAN, L. H.: The invertebrates. Vol. II. Platyhelminthes and rhynchocoela. The acoelomate bilateria, pp. 182—190. New York: McGraw-Hill Co. 1951.

JACOB, F., and J. MONOD: Genetic regulatory mechanisms in synthesis of proteins. J. molec. Biol. 3, 318—356 (1961).

KESSEL, R. G.: An electron microscope study of nuclear-cytoplasmic exchange in oocytes of *Ciona intestinalis*. J. Ultrastruct. Res. 15, 181—196 (1966).

LANG, P.: Über Regeneration bei Planarien. Arch. mikr. Anat. 79, 361—421 (1912).

LEBLOND, C. P., B. MESSIER, and B. KOPRIWA: Thymidine-H³ as a tool for the investigation of the renewal of cell populations. Lab. Invest. 8, 296—308 (1959).

LENDER, T.: Factors in morphogenesis of regenerating fresh-water planaria. In: Advances in Morphogenesis (Ed.: M. ABERCROMBIE and J. BRACHET), Vol. 2, pp. 305—331. New York: Academic Press Inc. 1962.

MANASEK, F. J.: Mitosis in developing cardiac muscle. J. Cell Biol., in press (1967).

NEEDHAM, A. E.: Regeneration in the Arthropod and its endocrine control. In: Regeneration in animals and related problems (Ed.: V. KIORTSIS and H. A. L. TRAMPUSCH), pp. 283—323. Amsterdam: North-Holland Publishing Co. 1965.

NICHOLAS, J. S.: Regeneration in vertebrates. In: Analysis of development (Ed.: B. H. WILLIER, P. A. WEISS, and V. HAMBURGER), pp. 674—698. New York: W. B. Saunders Co. 1955.

NORMAN, W. P., and A. J. SCHMIDT: The fine structure of limb tissues of the adult newt, *Diemictylus viridescens*. J. Morph. 123, 251—270 (1967).

OBERPRILLER, J.: A radioautographic analysis of the potency of blastema cells in the adult newt, *Diemictylus virideciens*. Growth 31, 251—296 (1967).

PUCK, T. T., and J. STEFFEN: Life cycle analysis of mammalian cells. I. A method for localizing metabolic events within the life cycle, and its application to the action of colcemide and sublethal doses of X-irradiation. Biophys. J. 3, 379—397 (1963).

REYER, R. W.: Regeneration in the amphibian eye. In: Regeneration (Ed.: D. RUDNICK), pp. 211—265. New York: Ronald Press Co. 1962.

ROSE, S. M.: Regeneration. In: Physiology of the amphibia (Ed.: J. A. MOORE), pp. 545—622. New York: Academic Press, Inc. 1964.

ROSE, F. S., and S. M. ROSE: The role of normal epidermis in recovery of regenerative ability in x-rayed limbs of *Triturus*. Growth 29, 361—393 (1965).

SALPETER, M. M., and M. SINGER: The fine structure of mesenchymatous cells in the regenerating forelimb of the adult newt. *Triturus*. Develop. Biol. 2, 516—534 (1960).

SCHMIDT, A. J.: The molecular basis of regeneration. Urbana: University Ill. Press 1966.

—, and M. WEARY: Localization of acid phosphatase in the regenerationg forelimb of the adult newt. *Diemictylus viridescens*. J. exp. Zool. 152, 101—114 (1963).

SCHOTTÉ, O. E.: The origin and morphogenetic potencies of regenerates. Growth Suppl., 59—76 (1940).

SINGER, M.: In: Regeneration in animals and related problems (Ed.: V. KIORTSIS, and H. A. L. TRAMPUSCH), pp. 375—376. Amsterdam: North-Holland Publishing Co. 1965.

Singer, M., A. Weinberg, and R. L. Sidman: A study of limb regeneration in the adult newt, *Triturus*. by infusion of solutions of dye and other substances directly into the growth. J. exp. Zool. **128**, 185—218 (1955).

Slautterback, D. B., and D. W. Fawcett: The development of the cnidoblasts of *Hydra*. An electron microscope study of cell differentiation. J. biophys. biochem. Cytol. **5**, 441—452 (1959).

Spallanzani, L.: Dissertions relative to the natural history of animals and vegetables. Translated from the Italian. London 1789.

Steen, T.: Differentiation of chondrocytes during limb regeneration in the Axolotl (*Siredon mexicanum*). PhD Thesis, Yale University 1967.

Steinberg, M. S.: Cell movement, rate of regeneration, and the axial gradient in *Tubularia*. Biol. Bull. **108**, 219—234 (1955).

Stephan-Dubois, F.: Les neoblastes dans la regeneration chez les planaires. In: Regeneration in animals and related problems (Ed.: V. Kiortsis and H. A. L. Trampusch), pp. 112—130. Amsterdam: North-Holland Publishing Co. 1965.

Steward, F. C.: The control of growth in plant cells. Sci. Amer. **209**, 104—113 (1963).

Takata, C., J. F. Albright, and T. Yamada: Lens antigens in a lens regenerating system studied by the immunofluorescent technique. Develop. Biol. **9**, 385—397 (1964).

Tardent, P.: Regeneration in the *Hydrozoa*. Biol. Rev. Biol. Rev. **38**, 293—333 (1963).

Thornton, C. S.: The histogenesis of muscle in the regenerating forelimb of larval *Amblystoma punctatum*. J. Morph. **62**, 17—47 (1938).

—, Studies on the origin of the regeneration blastema in *Triturus viridescens*. J. exp. Zool. **89**, 375—390 (1942).

Trampusch, H. A. L., and A. E. Harrebomée: Dedifferentiation a prerequisite of regeneration. In: Regeneration in animals and related problems. (Ed.: V. Kiortsis and H. A. L. Trampusch), pp. 341—374. Amsterdam: North-Holland Publishing Co. 1965.

Weiss, P.: Principles of development. New York: H. Holt and Co. 1939.

— Perspectives in the field of morphogenesis. Quart. Rev. Biol. **25**, 177—198 (1950).

Wolff, E.: Recent researches on the regeneration of *Planaria*. In: Regeneration (Ed.: D. Rudnick), 53—84. New York: Ronald Press Co. 1962.

Yamada, T.: Control of tissue specificity: the pattern of cellular synthetic activities in tissue transformation. Ann. Zool. **6**, 21—31 (1966).

—, and C. Takata: An autoradiographic study of protein synthesis in regenerative tissue transformation of iris into lens in the newt. Develop. Biol. **8**, 358—369 (1963).

The Acquisition, Maintenance, and Lability of the Differentiated State in Hydra

Allison L. Burnett

Developmental Biology Center, Department of Biology, Western Reserve University, Cleveland, Ohio 44106

I. Introduction

The title of this collection of papers, "The Stability of the Differentiated State", implies that the organisms discussed herein contain differentiated cells. However, one of the central questions we shall be discussing throughout the text is: does differentiation signify a fixed, irreversible state of a cell? For me, using *Hydra* as an experimental model, to accept the definition that a differentiated cell cannot dedifferentiate would be tantamount to saying that the animal probably contains no differentiated cells. If this is true then none of the material presented here will be relevant to a discussion of differentiation. There is one thing to be said, however, for the foregoing definition — it is a clear formulation and not subject to semantic bickering. All other definitions of differentiation are fraught with exceptions and lead invariably to long, often tedious hours of endless debate that leave symposium participants with ragged nerves, bad digestion and an urge to get back into the laboratory and get to work because the discussion accomplished nothing.

Thus, we have been living for a few decades now with a totalitarian pronouncement on the one hand and anarchistic pronouncements on the other because everyone, if pressed, seems to be able to come up with a definition which bests suits his research interests. The only way to eliminate the first definition is to show that every cell in the animal kingdom can dedifferentiate. Then we can securely say that there is no such thing as a differentiated cell. If we choose the anarchistic pronouncement and accept the fact that differentiation means just what each individual investigator says it means, then we must proceed with good-will hoping that pretty much all of us essentially agree on what we mean by differentiation and let it go at that. Most of us are, in fact, doing just this most of the time.

So without going into a definition let us say that in *Hydra* there are ten cell types which can be recognized from each other morphologically, physiologically, and chemically, with the exception of interstitial cells which can be categorized morphologically only by their *lack* of organelles when viewed at the level of the light microscope. It is to these cells I refer when I use the term differentiated. They are:

1. nerve cells (3 varieties)
2. cnidoblast cells (4 varieties)
3. mucous cells of gastrodermis (2 varieties)
4. gland cells of gastrodermis (1 variety)

5. interstitial cells
6. epithelio-muscular cells of epidermis
7. digestive cells of gastrodermis
8. mucous cells of basal disk
9. sperm cells
10. egg cells.

The following discussion will deal with means by which cells in adult *Hydra* achieve the differentiated state, how they maintain this state, and means whereby maintenance is destroyed and dedifferentiation results. Also, the question of redifferentiation will be raised and it will be clearly shown that a dedifferentiated cell can redifferentiate into a cell type other than that from which it was derived.

II. The Acquisition of the Differentiated State

It will be impossible to discuss all the cell types in *Hydra* in this paper, so let us employ the interstitial cell or neoblast as a case study. This small, basophilic cell is located in the epidermis and has been claimed by many authors, including myself, to constitute an embryonic reserve in the adult animal. There is reliable evidence that these auto-reproductive cells are the sole source of nerve cells, cnidoblast cells, sperm cells and *ova*. It has also been claimed that these cells can form other cell types as well, but for the present we will consider only epidermal cell types.

If we examine a normal *Hydra* histologically we find that the interstitial cell plus its derivatives are not distributed uniformly along the body column (Fig. 1). In the hypostome and tentacles interstitial cells are rare or absent. These areas are characterized by a ring of nerve cells running circularly at the bases of the tentacles and scattered nerve cells througout the length of the tentacle. A further search for nerve cells reveals that they are sparse in the underlying gastric region, and in some species they are extremely rare. However, at the junction of the peduncle with the budding region nerve cells again appear. They are distributed in concentrations nearly as high as in the tentacle throughout the peduncle and into the basal disk where they reach their highest concentration.

Upon observing this distribution one naturally asks why interstitial cells which enter the hypostomal region and peduncle differentiate into nerves whereas those in the gastric region are not forming nerves? An examination of the gastric region reveals that interstitial cells are performing two different functions. They are dividing to form nests of 8—16 cells. After nests have formed their enclosed cells differentiate simultaneously into cnidoblast cells which form nematocysts. Differentiating cnidoblasts are rare or absent in the tentacles, peduncle, and hypostome, nor are cell divisions in interstitial cells common in these areas.

It was this simple observation of the normal animal which convinced us that the explanation of polarity which we offered several years ago (BURNETT 1961) was relevant to an explanation of control of cell differentiation in regenerating animals. After all, polarity depends upon similar cells at the two cut surfaces pursuing different pathways of differentiation.

BURNETT (1966) presented a single model which interpreted both polarity and cell differentiation. Without going into detail we will present the essence of the model as it applies to cell differentiation.

It is proposed that cell differentiation in *Hydra* is controlled by a single inducer varying quantitatively from the hypostome to the base in an apico-base gradient. Quantitative differences in inducer result in qualitative differences as expressed at the level of cell differentiation. The action of the inducer is tempered by an inhibitor

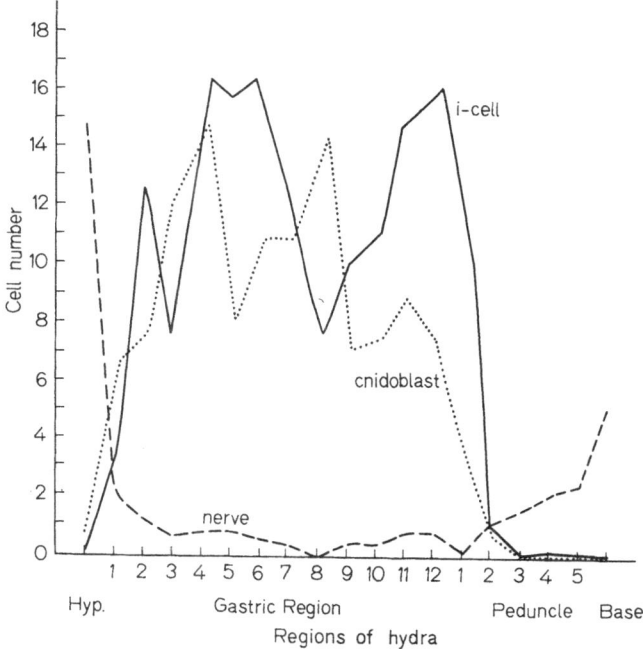

Fig. 1. Distribution of epidermal cell types along the length of the body column of *Hydra pseudoligactis*. Every epidermal cell type listed was counted on one half of seven longitudinal sections which passed through the mouth and the center of the base. Each morphological region of the animal was divided into equal sectors with the base and hypostome occupying one sector only. Only non-budding forms were counted in this study. Cnidoblast cells are only those cells forming nematocysts and do not include those with the fully formed organelle. The above graph represents counts from a single typical animal. i-cell = interstitial cell

produced by dividing cells in the gastric region beneath the hypostome. Therefore, it is the ratio of inhibitor and inducer in any body region which determines the fate of the cells in that region.

Using the interstitial cell as a case study we can more easily visualize the foregoing hypothesis. The + signs in the scheme below indicate the relative amounts of inducer and the subsequent pathways of differentiation of the interstitial cells.

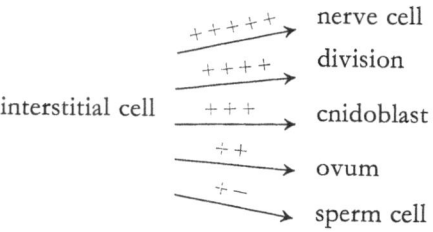

Consider the first three pathways of differentiation and relate them to the distribution of these cell types in the normal animal. Nerves are concentrated in the hypostome, cell division to form nests occurs in the gastric region and cnidoblasts are formed after a nest is formed. In this scheme inducer bound to interstitial cells becomes progressively reduced per cell during division. When the reduction of inducer is sufficient, then the next step on the hierarchy, differentiation into cnidoblasts occurs.

Evidence presented by LESH and BURNETT (1966), BURNETT, DIEHL, and DIEHL (1964), and BURNETT and DIEHL (1964) suggests that the inducer is a small peptide and is produced by nerves in the hypostomal region. If this is correct then the system becomes both a self-generating and self-limiting one. Interstitial cells forced by divisions into the hypostome will not divide further to form nests because they are in an area of high inducer which favors differentiation into nerve. These nerves will secrete further inducer insuring that nerve differentiation in this region will continue and provide a source of this cell type in the tentacles which are constantly sloughing cells at their extremity.

It would appear that increased nerve will result in increased inducer production, but when it is remembered that increased inducer will results in increased divisions in the gastric region thereby producing excess inhibitor, the system remains rigidly controlled. Divisions in the sub-hypostomal region provide continual binding sites for inducer issuing from the hypostome, but those nests located more proximally in the gastric column plus nests which have grown to a size that limits the passage of inducer to cells on the interior will differentiate into cnidoblasts.

Before discussing germinal differentiation in terms of the present differentiation hierarchy, it is necessary to consider a final question concerning distribution of cell types in the normal animal. Why do interstitial cells upon entering the peduncle differentiate once more into nerves? The explanation for this is found in observing the difference in diffusion gradients of inducer and inhibitor. Inhibitor, unlike inducer, is not bound to cells but leaks readily to the culture medium. Culture medium in which animals are grown in crowded conditions under aseptic conditions contains an inhibitor which will completely inhibit budding and regeneration and under decreasing concentration will interrupt these processes in various stages (TARDENT 1960; ROSE and ROSE 1941; DAVIS 1966; BURNETT 1966; LENIQUE and LUNDBLAD 1966; BURNETT and RUFFING 1967).

Since the inhibitor is a leaky molecule it does not extend in a gradient from the hypostome to the base but extends only about half way down the body column (Fig. 2). This results in the formation of an area where residual bound inducer does not compete with inhibitor. In other words, the ratio of inducer and inhibitor is favorable once again for nerve differentiation. This area in *Hydra* is at the junction of the peduncle and budding region. This brings about a reversal of the phenomenon witnessed in the hypostome. Just distal to the area where nerve differentiation begins (that is where inhibitor is dilute but still present) the next step on the differentiation hierarchy, cell division begins at a high rate. This results in the initiation of the bud which arises just distal to the peduncle.

When the bud begins to grow additional inhibitor also forms and this diffuses both basally and distally in the parent column. In fact, if the parent column is excised in the midgastric region while a bud is forming, it will not regenerate until the bud has formed at base and is ready for detachment. This equalizes inhibitor throughout the gastric region (Fig. 3), a suggestion that has been overlooked unfortunately by many investigators in this area who traditionally employ budding forms to elaborate growth patterns in non-budding animals.

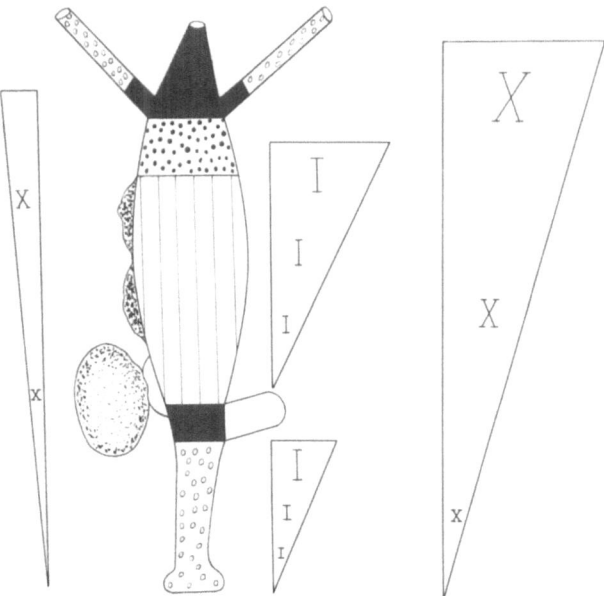

Fig. 2. This model suggests how the inducer-inhibitor ratio controls cell differentiation in the various body regions of *Hydra*. In the extremities (open circles), no cell divisions occur and interstitial cells are rare or lacking. Dark areas (hypostome and junctions of budding region and peduncle) indicate where interstitial cells differentiate into nerve. This can only occur where there is excess inducer (X) in proportion to inhibitor (I). Solid circles show the growth region which, because of its proximity to the hypostome, is an area high in cell division. The gastric region (lined area) is where interstitial cells complete their divisions forming nests of several cells which then differentiate into cnidoblasts. Those interstitial cells which do not differentiate into cnidoblasts in the gastric region differentiate into nerves at the junction of the budding region and peduncle. On the left-hand side of the model is the gradient of inducer (X) that occurs during sexuality. Inducer production is reduced or ceases entirely. Interstitial cells in the presence of reduced inducer concentration differentiate into sperm (testes indicated as dotted protuberances on left side of animal) and in areas where inducer is lacking or present in trace amounts into ova (below testes at level of bud)

In Fig 3 inhibitor is pictured as diffusing only basally. At the area where it becomes sufficiently dilute one would expect another area of growth. However, at this level stimulator is also dilute or absent. This results in the formation of the second developmental realization of the *Hydra* body column, a base.

Let us now consider what occurs when neurosecretion in the nerve neurones ceases. There is evidence that in *Hydra* neurosecretion may be controlled by temperature, pCO_2 and even day length depending upon the species of animal employed. Cessation of neurosecretion results eventually in a cessation of cell divisions in epithelial cells (lack of inhibitor production) and cessation of all somatic differentiation on the part of interstitial cells. Some species enter "partial" sexuality where they may continue to bud while forming gonads. During complete sexuality every interstitial cell in the animal differentiates germinally and budding ceases (BURNETT and DIEHL 1964).

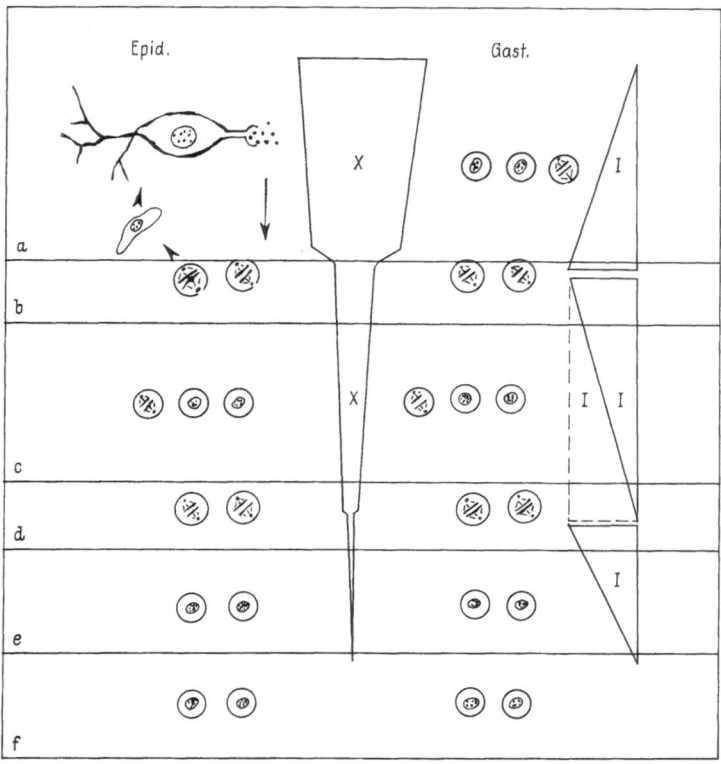

Fig. 3. Model depicting the uptake of inducer and the release of inhibitor by dividing cells in the various body regions of *Hydra*. a Hypostome; b Growth region; c Gastric region; d Budding region; e Peduncle; f Base. Inducer (X gradient) is produced by nerves in the epidermis of the hypostome and is bound by adjacent cells. The largest uptake is in the growth region and budding region where the frequent divisions continually provide new binding sites. Thus, the gradient is not a uniform one but falls off sharply in the growth and budding region. The inhibitor (I) is produced in areas where cell divisions occur. No inhibitor is formed in the peduncle and basal disk but diffuses into this region from the adjacent budding region. Dotted lines indicate that, during the time of asexual reproduction, inhibitor actually diffuses up the body column as well as proximally. This accounts for the relative uniformity of divisions in the gastric region where inhibitor from the growth and budding regions combines

During partial sexuality inducer can still be extracted from animals but in reduced amounts; during complete sexuality no inducer can be extracted. As seen in Fig. 2 when the inducer gradient becomes progressively diminished somatic differentiation ceases and sperm are formed. When inducer is completely lacking or present in trace amounts ova are formed. We have been able to induce ova formation in five male cultures of species formerly considered dioecious.

Although other cell types show changes similar to interstitial cells depending upon where they exist in an inducer gradient space forbids a discussion of these changes. The reader is referred to the paper in which the model is presented in more detail (BURNETT 1966).

In summary, we can hypothesize that one factor which controls the acquisition of the differentiated state is *the position of the cell in an inducer gradient*. Later we shall see that the presence of inducer is not only necessary for differentiation but necessary as well for the maintenance of the differentiated state.

III. The Maintenance of the Differentiated State

The question of maintenance of differentiated cell lines is probably in some respects more important than events leading towards differentiation. There is general agreement today that all cells of an organism contain qualitatively identical genomes and diversity in cell populations is an indication that different genes are functioning in different cell types. Most research effort in development today is an attempt to discover the mechanism through which gene action is controlled. We must decide whether this control is genetic, that is, does a cell pass an inherited capacity for a particular type of differentiation to its daughter cell, or whether other controlling mechanisms exist.

Since we have already stated that in *Hydra* a single inducer controls the direction of differentiation of the neoblasts, let us look more deeply into animals in which inducer is diminished or lacking. What changes occur in somatic cells? Are these changes reversed if additional inducer is added?

Before discussing interstitial cells and their derivatives let us consider epithelio-muscular cells of the epidermis. These cells constitute the bulk of the epidermis and extend from the basal disk to the tips of the tentacles. How are they affected by their position in the inducer gradient?

Fig. 4 traces the passage of an epidermal cell down the column from the sub-tentacular growth zone. Epithelial cells divide more frequently in the region just below the hypostome than they do in the hypostome itself or in the mid-gastric region. Thus, epidermal cells are forced continually down the column from the sub-tentacular region. The fully differentiated epidermal cell is characterized by an external mucous secretion (muco- or glycoprotein, not found elsewhere in *Hydra*), and the presence of a contractile longitudinal myoneme whose contraction shortens the animal. The evidence is strong that these cells also contribute to the formation of a collagen-like mesoglea.

When the epidermal cells reach the budding region their mitotic activity is increased again. This is not obvious from the figures recorded in Fig. 5, because the size of the fields counted were large and the small area of intense cell division just proximal and distal to the bud on the parent column was "diluted" in the over-all tally. However, Fig. 6 where every dividing cell in the hydra was counted by SONDRA CORFF in our laboratory, the regions of growth are more obvious. Epidermal cells upon entering the peduncle have a drastically reduced division rate. Practically all divisions that occur are in the upper $1/3$ region of the peduncle and few or no divisions are seen in the proximal $2/3$ region.

8*

However, as the epidermal cells enter the region of the basal disk they enter a second differentiated state. The secretion of the external muco- or glycoprotein

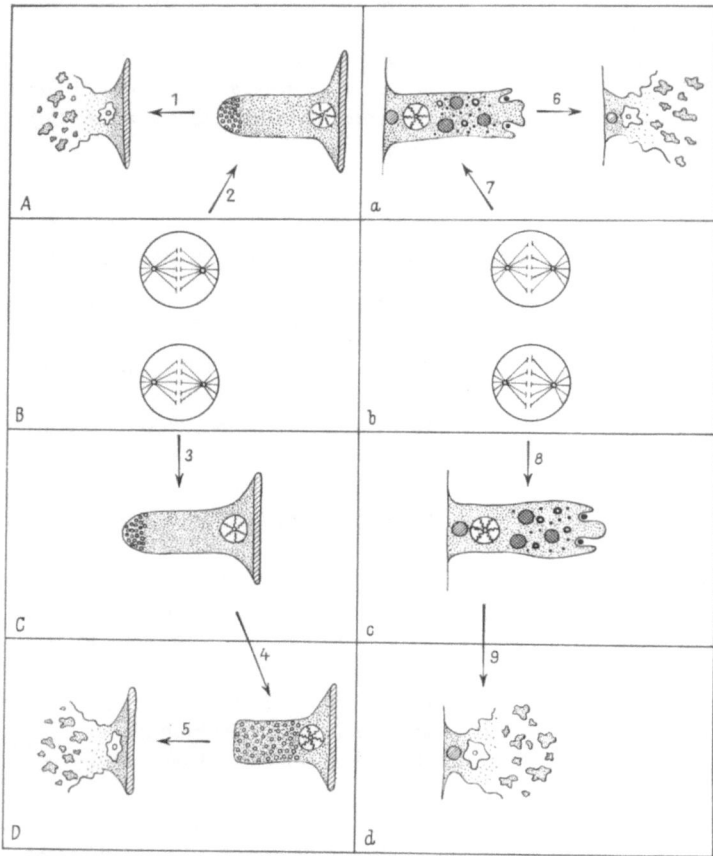

Fig. 4. Schematic representation of the fate of epithelio-muscular cells (left side of diagram) and digestive cells (right). Blocks represent different body regions of *Hydra*. A Tentacles; B Growth Region; C Gastric Region; D Basal Disk. Epithelio-muscular cells in the growth region enter divisions more frequently than in adjacent regions. Cells forced distally into the tentacles cease division (2) and differentiate by secreting a mucous border and forming an intracellular myoneme. When these cells reach the tips of the tentacles (1) they die and are sloughed off the column. An epithelio-muscular cell forced proximally also differentiates a mucus border and myoneme (3), but before death at the level of the basal disk (5) the cell secretes an acid-mucopolysaccharide (4) not found elsewhere in the epidermis. Digestive cells forced distally from the growth region (7) engulf large quantities of food, cease division, and form an intracellular myoneme that runs circularly around the body column. At the tip of the tentacles the digestive cells (6) are sloughed into the enteron. A similar cell forced proximally from the growth region undergoes differentiation similar to cells forced distally (8) and is finally eliminated (9) at the level of the basal disk

ceases and the cells form an acid mucopolysaccharide. This material upon liberation from the cell provides for the attachment of *Hydra* to a substratum (see Philpott, Chaet, and Burnett 1966).

Thus, an epidermal epithelio-muscular cell has three developmental pathways open to it. It may divide, it may form a longitudinal myoneme and an external muco- or glycoprotein border, or it may differentiate into a mucous cell containing an acid mucopolysaccharide. Each of these pathways may be correlated with a morphological region where one pathway is favored over another.

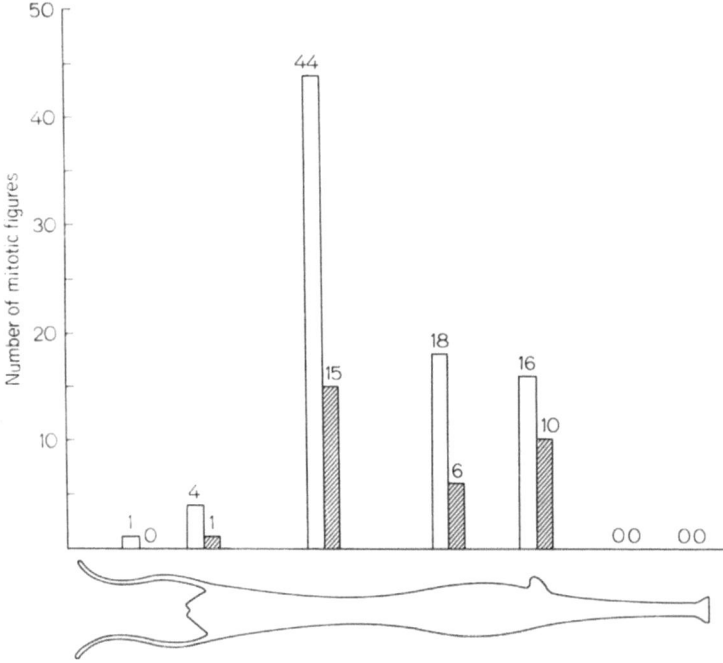

Fig. 5. Number of mitotic figures observed in epitheliomuscular and gastrodermal digestive cells in 8 *Hydra pseudoligactis*. 4,200 cells were counted in each animal for a total count of 33,600 cells. Only metaphase and anaphase stages are included here in order to insure that the cells actually were in division at the time of fixation. The various body regions recorded are the tentacles, the hypostome, the sub-hypostomal growth region, the gastric region, the budding region, the peduncle and basal disk. Open bars: epidermis; hatched bars: gastrodermis.

If, as we have assumed, a hypostomal factor controls these pathways by varying quantitatively along the body column, what is the fate of these epidermal cells when secretion of the hypostomal factor ceases? This situation is obtained in the animal undergoing advance sexuality. In *Hydra pirardi* the axons of the nerve cells recede (BURNETT and DIEHL 1964) and no active inducer fraction can be collected (LESH and BURNETT 1966). Histological examination of these animals reveals that the mucous cells of the base are no longer limited to the most proximal region of the animal. Instead there is a shift in the distal direction whereby the epithelial cells of the pe- duncle transform into mucous cells. Thus, the original base of the animal has now moved up the budding region. Similar cells of the budding region and gastrodermis now take on the morphology of peduncular cells, i.e., they are characterized by a lack of basophilia and contain large vacuoles. Cell divisions throughout the gastric region cease.

Therefore, as the gradient of inducer which normally extends from the hypostome to the base shortens in a proximodistal direction, the epidermal cells, almost as if they were "tailing" the shortening gradient differentiate accordingly. This recession of the gradient is witnessed dramatically as mucous cells characteristic of the base form more and more distally along the column.

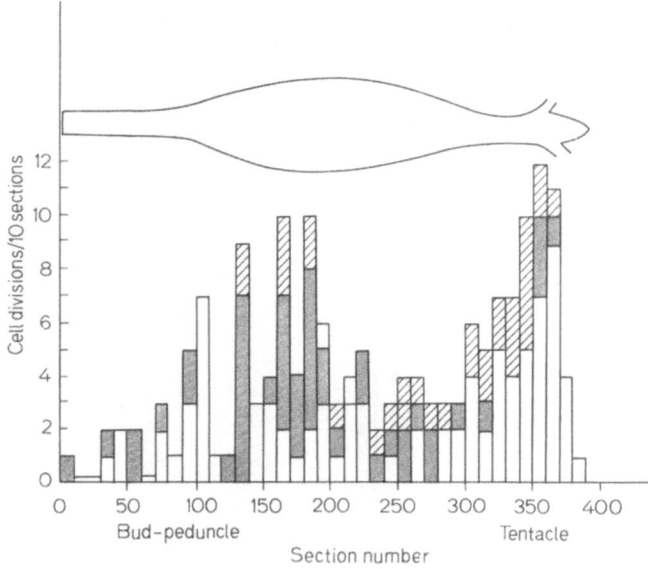

Fig. 6. Distribution of mitotic figures along the body column of *Hydra oligactis*. Every cell in the animal was counted. In this graph only those cells which influence net movement of tissue layers down the body column are recorded. Interestingly, if every tenth section were counted and the number of divisions multiplied by ten and then plotted (method used by CAMPBELL (1966) see text) the increase in mitotic activity seen in the budding region would be shunted forward and appear as a peak in the gastric region. The above graph represents counts from a single animal. Shifts in the peaks will occur during budding, growth from bud to adult, etc. In this animal the lack of divisions at the extremities of the column plus the decrease in divisions in the middle gastric region are evident. Open bars: epidermal epithelio muscle cells; dark bars: digestive cells; hatched bars: gland cells

As already noted, the interstitial cells during this period no longer differentiate in a somatic direction. The entire population of cells differentiates into gametes. In *Hydra pirardi* as the inducer gradient recedes the interstitial cells differentiate into sperm cells at first, but in proximal regions (the site of the original budding region) as soon as no inducer is present ova are formed. This sequence of differentiation first into sperm and finally into ova is exactly what is predicted by the model presented earlier. Also, it is interesting to note that in hermaphroditic animals such as *Hydra viridis*, testes always form distally and ovaries proximally. This again represents differentiation in a receding gradient.

Also, it is of interest to describe the morphological changes which occur when inducer production ceases. The tentacles become increasingly shorter until they finally appear only as tiny knobs tightly surrounding the hypostome. In more advanced sexuality the tentacles, plus the mouth and hypostome disappear altogether. This results in an animal which is essentially a two-layered sac covered with testes, and more proximal regions, a few ovaries. To the general observer the animal would never be recognized as a *Hydra*.

Animals at this advanced stage of sexuality will invariably perish, but they may be recovered in several ways. If the animals are brought back to room temperature slowly, five or six animals out of one hundred may recover. These are animals which still contain a few interstitial cells which have not yet differentiated into gametes. Also it is possible to graft on a normal hypostome so that a fresh supply of inducer plus a few interstitial cells are introduced into the sexual forms. These animals when brought back to room temperature show a good recovery rate. In the first case where no graft is made the animal had to be fed by injecting fresh *Artemia* extract directly into the gastrovascular cavity, because these forms possess no tentacles or nemato-cytes to subdue prey.

In forms which recover without grafting it is interesting to note that the first change noticeable when they are brought to room temperature is the appearance of axons from the rounded nerve cell bodies in the distal region. Once nerve is reformed and neurosecretion resumes the remaining interstitial cells begin to divide again or differentiate in a somatic direction. Similarly, digestive cells and epidermal epithelio-muscular cells begin to divide. Once growth is resumed tentacle buds are formed and a new hypostome elaborated. As epithelial cells continue to divide the extensive region of mucous cells is forced basally. Within two weeks the differentiation se-quence of epithelial cells and interstitial cells is completely normal and the animal cannot be distinguished morphologically from animals maintained in the culture dishes at room temperature.

We conclude that the steady production of inducer is necessary not only for the control of differentiation but is also necessary for the animal to maintain a normal form. Recently, CAMPBELL (1966) and CLARKSON and WOLPERT (1967) have questioned whether growth plays any part in the achievement of form. In the present analysis it is seen that cell division is an important step in the differentiation hierarchy and one which is absolutely necessary for the maintenance and the achievement of form.

Also, it is important to mention here that although nerve cells are the source of inducer, they also require the feed-back of this inducer on themselves in order to function. In animals placed at low temperature it will be remembered that axons recede and the nerve cell body rounds up into a sphere. If the hypostome from a nor-mal animal of the same or even a different species is grafted onto a sexual form it will continue to produce inducer for a few days even if host and graft are left at low temperature. During this time, in the presence of inducer supplied from the graft, the rounded nerve cell bodies form new axons and cannot be distinguished from normal nerves.

IV. The Lability of the Differentiated State

Already we have seen that the differentiated morphological state of *Hydra* is drastically altered in the absence of a hypostomal inducer. At the cellular level we have discussed this phenomenon in terms of interstitial cell and epidermal cell differ-entiation. In the following sections we will concern ourselves with morphologically and physiologically differentiated cell types and examine them in light of the following questions: 1) Can a differentiated cell type dedifferentiate (morphologically and physiologically) and the resultant cell redifferentiate into a cell type other than that from which it was formed? 2) Does a differentiated cell after division pass on the capacity for a specific type of differentiation to its daughter cells?

A few years ago we found that *Hydra viridis* can be regenerated from an explant consisting of only two cell types (see HAYNES and BURNETT 1963; DAVIS et al. 1966; and BURNETT 1966). This is accomplished by separating the cell layers of *Hydra viridis* and isolating small portions of the gastrodermis of the gastric region. This region consists of only digestive cells and gland cells (the mucous cells of the gastrodermis are confined to the hypostome of this species). These explants are then transferred to an ionic medium which preserves the integrity of the cells (they quickly rupture and die in normal culture medium). It must be stressed that this ionic medium is not one which is merely iso-osmotic with the explant, but the choice of anions and cations are equally important. Moreover, this medium, although it will support the existence of a gastrodermal explant, will not support an intact animal.

During the regeneration process the explant is transferred three times through a diluted series of ions until it is finally placed in culture water. During various stages of regeneration the explant was examined histochemically and with the electron microscope.

Only those cells at the exterior of the explant, that is, those in direct contact with the ionic medium underwent change. Gland cells at the periphery shed their enclosed secretory granules and lost their well developed endoplasmic reticulum. Without dividing, they reached a stage where they could not be distinguished ultrastructurally from interstitial cells. These cells then divided to form nests and thereafter differentiated into cnidoblast cells which formed functional nematocysts. The interstitial cells in the area of the animal destined to form the hypostomal region differentiated into nerve cells. Animals cloned from a two cell regenerate were able to form normal sperm cells from their interstitial cells (BURNETT, DAVIS, and RUFFING 1966).

Digestive cells on the periphery of the explant broke down their enclosed algal bodies, elaborated longitudinal muscle fibers, and began secreting the mucoprotein border characteristic of epidermal cells. Gastrodermal digestive cells normally never form this mucous product at any level of the body column. The transformation from digestive cells to epidermal cells is a direct one and need not involve cell division.

These transformations represent two distinct types of metaplasia. In one case the cell reverts to a neoblast type cell which then divides and then differentiates into a cell type other than that from which it was derived. In the transformation of digestive cells the cells do not revert to an embryonic type state, do not undergo divisions, and by forming new muscles, mucous, and possibly mesoglea, transform directly into epidermal epithelio-muscular cells.

These results demonstrate that a differentiated cell type in *Hydra* can undergo a complete dedifferentiation and redifferentiation if treated with the proper ionic media. In the normal animal when a gland cell divides it retains its secretory product even during division and thus the daughter cells retain their differentiated capacity. Thus, at this point we would be forced to say that the differentiated state can be an inherited one only if the proper environment is available. If this environment is altered properly the commitment to the original differentiated state is lost and the cell can pursue a new pathway of differentiation. Also, these results show that this dedifferentiation sequence is not limited only to somatic cell types, but that a somatic cell placed in the proper environment can be induced to undergo meiotic divisions.

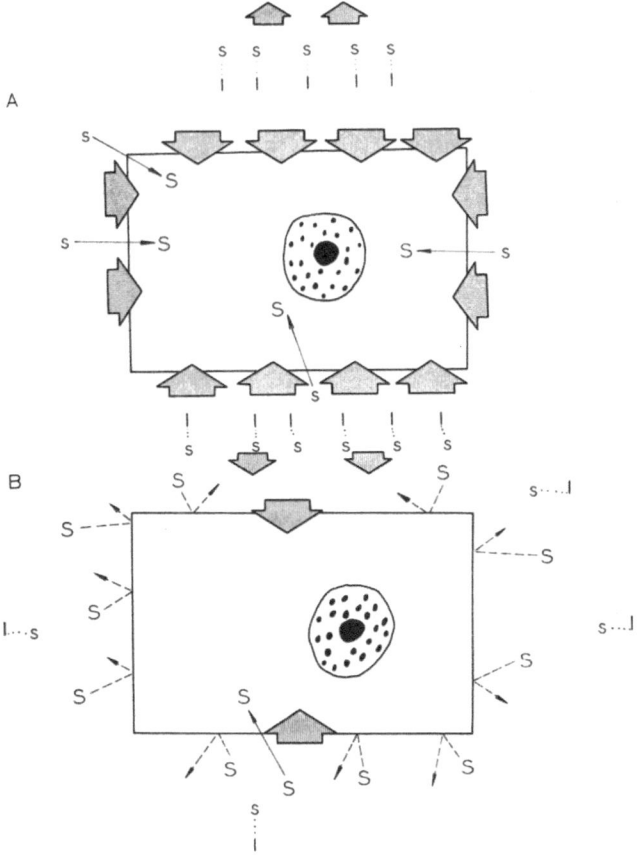

Fig. 7. Model suggesting how inducer and inhibitor may act at the cellular level. The inducer (large arrows) becomes associated with the cell membrane and alters permeability. The inhibitor (I) attaches to a substrate (S) in intercellular spaces. When the inhibitor leaks to the culture medium, it carries the attached substrate with it. In 7 A the inducer and inhibitor are concentrated around a cell. Although substrate is withdrawn in large amounts by the inhibitor, the cell is still an efficient "gatherer" of substrate because of the large amount of inducer present. Thus, substrate (S) enters the cell. In 7 B, although substrate is available, little of it can enter the cell because of scarcity of inducer. Thus, the inducer-inhibitor ratio determines the amount of substrate which will enter the cell

The reader may question why ions are so important in controlling the fate of differentiation, especially since we have previously invoked a single inducer to explain this phenomenon. For reasons which we will not go into here, it was suspected that the inducer might act by changing the permeability of the cells to ions or small substrates. Thus, the amount of inducer bound to a cell would largely determine the ionic milieu of the cell's interior. Since inhibitor does not become bound to inducer or to the cells, we postulated that while the inducer makes a cell more efficient in accumulating ions, inhibitor in leaking from the animal becomes bound to ions and removes them from the animal. Thus, the ratio of inducer and inhibitor would determine both the internal and external ionic milieu of the cell. Fig. 7 schematically represents this hypothesis.

We have found that by varying inducer quantitatively or by adding inhibitor to the animal the fate of interstitial cells can be controlled (BURNETT 1966, and unpublished results). Interstitial cells in the gastric region can be induced to form nerve cells by the hundreds, they can be stimulated to remain in a steady dividing state without differentiation (nests of over 20,000 cells can be induced), they can be induced as a total population to form cnidoblasts, and finally can be induced as total populations to form sperm or ova.

If this induction does result from the effect of inducer on cell permeability, it would appear that similar changes might be stimulated by forcing ions into the animal by increasing their concentration in the external culture medium. A second's reflection will remind the reader that under normal conditions the epidermal and gastrodermal cells share different environments both with respect to inducer-inhibitor ratio and external ionic milieu. Epidermal cells will eliminate their inhibitor directly to the culture medium or towards the gastrodermal cells. Gastrodermal cells, on the other hand, can liberate inhibitor only into the gastrovascular cavity where it will accumulate, or towards the epidermis. The same is true for ions eliminated from the cell layers. This observation may help in interpreting the results of the last experiment where only gastrodermal cells at the periphery of the explant (those that could leak directly into the external medium) were involved in transformation to epidermal cell types.

Whole animals, gastric annuli, and cells in tissue culture (to be discussed later) have been subjected to media of varying ionic concentration for various periods of time. A few observations taken from *Hydra pseudoligactis*, after such treatment will be cited here (see MACKLIN and BURNETT 1966).

Whole live animals incubated in high concentrations of $NaCl$ or $NaHCO_3$ exhibit a dramatic change in stainability with toluidine blue at pH 8 of nuclei after alcohol fixation. The nuclei instead of staining orthochromatically stain a bright pink. The reason for this metachromatic staining is not known, but it is possible that the high sodium concentration in the cells uncouples the DNA-histone combination just as it does in *in vitro* situations. This might presumably expose anionic groups in the nucleic acid, making it possible to form the toluidine blue polymer necessary for metachromatic staining. The cytoplasm of the interstitial cells and gland cells continues to stain orthochromatically.

The above observation takes on significance when $CaCl_2$ (0.2%) is added to the medium along with the NaCl. Within 12–24 hours every interstitial cell in the animal has differentiated into a cnidoblast and every gland cell of the gastrodermis has differentiated into a mucous cell. No basophilic staining is found in any cell. Thus, the entire pathway of differentiation of two cell types has been directed by a change in the ionic concentration of the external milieu. Similar changes in other cell types have been observed by altering the ionic milieu. Since we have been able to control cell populations in a similar fashion by altering stimulator-inhibitor ratios (BURNETT 1966), it is possible that control of differentiation is largely reflected in the state of permeability of the cell membrane.

How ions act in controlling differentiation is unknown. It is often suggested to the author that anything so unspecific as an ion could not possibly direct the activity of a genome, which specifies a particular cell type. However, it must be realized that in any organism there are relatively few pathways of differentiation open to a genome. The genome of *Hydra* will not specify, under any conditions, planarian cells. Ionic or substrate concentrations, even varying over wide ranges, may well be specific enough to "unmask" the genes which will specify enzymes necessary to make a specific cell product, mucous, for example. Other genes may be "unmasked" as well in the process, but these we may not detect because only the right combination of "unmasked" genes will result in the production of the end product by which the cell is recognized.

Although other changes in the differentiation sequence have been observed after treatment with ions (four-fold increase in division rate with potassium, cessation of mucoprotein production instead of acid-mucopolysaccharide in the basal mucous cells, that is, shifting the gradient in a distal-proximal direction) we have come to this general hypothesis: In order to cause a differentiated cell to dedifferentiate or to differentiate into another cell type, it is necessary to first change the commitment of the genome. This could presumably be accomplished by supressing the genes active in specifying enzymes for a particular pathway of differentiation, or by activating other genes hitherto repressed.

Commonly when it is suggested that environment determines the fate of a cell, it is argued that two cell types may be placed into the same culture medium and both retain their differentiated state; also their progeny retains the differentiation pathway of the parent. To this we reply that both cell types entered the common culture medium with genomes already committed for a particular synthesis. Both these genomes remained functional in the medium (which in most cases is a so-called "complete" medium in any case).

In our experience, we find that the cells have to be placed in a medium which will sustain them but simultaneously traumatize them. Although it is not known why Na^+ sensitizes a cell for the induction of a specific differentiation after the addition of Ca^{++}, it is known that after Na^+ treatment many ionic groups hitherto unexposed now appear in the nucleus. To our knowledge the only molecule in the nucleus capable of providing anionic groups in the regular spacing needed for metachromatic staining is DNA. The appearance of anions would depend upon DNA-histone uncoupling. The fact that NaCl is used to cause this uncoupling *in vitro* is interesting in this light. Also interesting is the fact that the ionic medium which supports dedifferentiation and redifferentiation of gastrodermal explants depends not solely on osmotic relations between cell and environment but upon specific cations and anions present in the medium. For example, if $MgSO_4$ is substituted for $MgCl_2$ dedifferentiation does not occur.

For the present, for reasons we hope will become clearer later, we reject the claims of workers who proclaim the fixity of cell lines. Their evidence for the fixity is mainly based on negative results. It must be remembered that vertebrates and many invertebrates have had millions of years to elaborate complex internal environments. Most attempts by investigators until now have been to duplicate these environments rather than determining the environment present when the uncommitted cell first attained its committed state. Diverse cell populations placed in identical media must first be "equalized" in terms of their genetic commitment and then restimulated to pursue a common pathway of differentiation.

Further interesting observations may be made upon entire populations of interstitial cells which are stimulated to differentiate in a common direction. There has
been a long controversy concerning the antagonism between growth and differentiation. It has often been stated that dividing cell populations are not differentiating
populations and only after divisions cease can differentiation begin.

In normal *Hydra* this is also the case. For example, an interstitial cell will divide
to form a nest. Once the nest of 8–16 cells has been formed, divisions cease and
differentiation to cnidoblasts begins. One may ask, whether divisions are a requisite
for differentiation. According to our scheme we would answer in the affirmative as
far as the normal animal is concerned. Let's suppose an interstitial cell has bound
1000 molecules of inducer. At each division the number of bound molecules of
inducer per cell is halved. New inducer may become bound to the dividing nest, but
as the cells are dividing they are progressively being forced lower on the inducer
gradient. Finally a stage is reached when the inducer concentration per cell has been
reduced to the level which favors the next step on the hierarchy, that is, differentiation
to cnidoblasts. A few interstitial cells in the nest (especially those at the periphery
exposed to additional inducer) may contain enough inducer to "escape" the differentiation process. In fact, these cells probably supply a continual reserve, otherwise all
interstitial cells would differentiate and be used up.

However, when the proper ions are *forced* into the animal all interstitial cells are
in a sense "equalized". As a result, all interstitial cells whether they have been destined
to divide, to form nerve, or whether they exist individually or in nests, participate
in cnidoblast formation. Thus, a cnidoblast cell does not *have* to arise from a cell
which has undergone previous divisions. It's just that in the normal animal this is
usually the case.

V. Determination and Differentiation *in vitro*

The biggest problem in developing an *in vitro* system which will support growth
and differentiation of cells is finding the proper medium. For the higher animals, at
least, this usually means culturing the cells in an environment which is borrowed
directly from the living environment in which the cells normally exist. As a result,
most tissue culture media are chemically not defined. Once serum and protein are
introduced into tissue culture, not only are the large molecules "unidentified" but
the ionic environment is similarly not known. Environments such as these have been
useful for solving certain problems concerning differentiation, but have been un-
successful systems in which to determine factors controlling gene action. Hydroids
appeared to be animals of choice for the development of known tissue culture media
because the animals are largely exposed to their natural external environment (usually
sea-water, – *Hydra* is an exception) and do not possess a circulatory system containing
proteins, unknown small substrates, hormones, ions, etc.

In collaboration with Dr. NECCO at the Stazione Zoologica we cultured *Tubularia*
stem tissue in a medium originally developed by Dr. NECCO and altered for our
experimental organism. This medium is chemically completely defined. It consists of
inorganic salts and contains only fumarate, pyruvate, and glutamine as organic
substrates. This medium supports growth and differentiation of both epidermal and
gastrodermal cell types. We have cultured cells for 15 days in roller tube preparations,

but in most experiments cells were cultured in hanging drops for six days. During this time all developmental events interesting to us were completed.

An explant from the distal stem region just below the hydranth of *Tubularia* when placed in hanging drop cultures shows the following behavior. The explant attaches in four hours by filopods extending from gastrodermal digestive cells. The gastrodermal digestive cells spread as a monolayer during the next twenty-four hours. Next the monolayer is invaded by interstitial cells and the epidermis which slowly spreads over its surface.

Interstitial cells entering the monolayer first divide to form nests of 8—16 cells. These nests then differentiate into cnidoblast cells which form functional nematocysts. Interstitial cells in a nest which "escape" cnidoblast differentiation continue to divide and form additional nests. Some interstitial cells entering the monolayer do not divide but differentiate directly into neurons. This is especially noticeable at the periphery of the original monolayer.

Digestive cells of the gastrodermis and epithelio-muscular cells of the epidermis divide normally and their progeny inherits directly the phenotype of the parent cell. Large basophilic cells, typically called basal reserve cells, differentiate in the gastrodermis to form huge gland cells which may measure 50—100 microns in length.

In short, every cell which divides in the normal animal divides in culture, and every event observed in the normal differentiation sequence of interstitial cells is observed as well. No cells dedifferentiate morphologically and all cell types are readily recognized after 15 days in culture. Thus, this medium may be considered "complete" at least for the time period the cells were under observation.

Interesting changes are observed if the external milieu of the cells is altered. Employing a wide range of Na^+ and Ca^{++} combinations, it was found that if the medium contained 5 parts Na^+ to 1 part Ca^{++}, *all* interstitial cells in the culture, whether they resided in the original explant or had invaded the monolayer, differentiated in 12 of 20 cultures into cnidoblasts. In the remaining 8 cultures, for reasons which probably reflect the physiological state of the original explant, the cells either became detached from the glass or began to slough cells from the periphery. If the Na^+-Ca^{++} ratio is 6:1 the cells retain their morphological identity and cannot be distinguished from normal cultures.

Again, we see that a simple change in the ionic milieu surrounding the cells induces a specific pathway of differentiation. It is interesting to note that the Na^+-Ca^{++} ratio here is different from that needed to induce differentiation in *Hydra*. In fact, it is different for different species of *Hydra*.

We continue to stress the importance of ions in maintaining a differentiated state plus their importance in determining the direction of differentiation of cells. As in experiments reported in the previous section, Na^+ alone is not sufficient to induce differentiation. For reasons not known, Ca^{++} is also necessary.

Once hydroid cultures are established in our laboratory in Cleveland we will test other ionic combinations plus the effect of different small substrates such as succinate and malate. Also, we will test the effect of inducer on the *in vitro* system. Preliminary experiments in Naples have revealed that there is a factor in the hydranth of *Tubularia* which also controls differentiation and growth.

Inducer was collected in the same manner as that described by LESH and BURNETT (1966) for *Hydra*. However, thin layer chromatography analysis showed our extract to consist of two proteins or polypeptides instead of one. This extract, prepared in normal culture medium, was added to twenty cultures after the monolayer had formed. Eleven of the cultures behaved similarly to those treated with ions, that is, all interstitial cells differentiated into cnidoblasts. In the remaining nine cultures, cell divisions were increased enormously. Although more quantitation remains to be done, nests of up to 60 interstitial cells were seen in simultaneous division. These divisions would produce nests of 120 cells, a situation never obtained in normal culture nor in the intact animal. Data from interstitial cell counts indicate the division rate was increased 5- or 6-fold.

It is interesting that different cultures responded in one of two ways to addition of inducer. In one case differentiation is induced; in another, growth. Again, the reason for this difference of response is not known, but it must be remembered that all animals used in these experiments were collected daily from the Bay of Naples. We have not determined the level of inducer already present in the explant nor do we know the physiological state of the animals observed. For example, were they just at the point of autotomizing their hydrants; were they continuing in a sexual state, or ending sexuality? Only when reliable conditions for the long-term culture of *Tubularia* are developed, these answers will be known.

We would rather adopt the positive attitude for the present and state that both a factor in the hydranth, plus the proper Na^+-Ca^{++} ratio are capable of inducing differentiation of interstitial cells, and in the former case, capable of inducing cell division. These observations are in keeping with results concerning *Hydra*, as discussed in the preceeding sections.

Acknowledgements

This work was supported by a grant from the National Institutes of Health (GM-11218-05). The author is also thankful for support from a Career Developmental Award (HD-00669-08) which allowed him to work for one year at the Stazione Zoological in Naples, Italy.

The author would especially like to thank Miss FAITH RUFFING and Miss JUNE ZONGKER for their very capable assistance and their development of hanging drop cultures for *Tubularia*. Also the author is indebted to Dr. PETER DOHRN, director of the Stazione Zoologica, and Zoology director, ANDREW PACKARD for their kindness and co-operation during the author's year at the Stazione Zoologica.

VI. Conclusion and Summary

Differentiated cell types in *Hydra* such as the gland cell and the digestive cell are autoreproductive and their phenotype is inherited by their progeny after division. However, if these cells are exposed to a controlled ionic environment consisting of the proper anions and cations, gland cells dedifferentiate to interstitial cells. These interstitial cells can then redifferentiate into cnidoblasts, nerves, or gametes. Digestive cells in this environment differentiate directly into epidermal epithelio-muscular cells.

Under normal conditions the achievement of the differentiated state and the maintenance of this state depends upon the presence of an inducer produced by neurosecretory cells in the hypostomal region. It is hypothesized that this single inducer controls all cell differentiation in the *Hydra* body column. The inducer varying quantitatively in an apico-basal direction produces qualitative effects as witnessed in the direction of cell differentiation. The action of the inducer is regulated by an inhibitor believed to be produced by dividing cells. Thus, the ratio of inducer and inhibitor at any body level determines the sequence of differentiation of cells at this level. Since ions have been shown to control cell differentiation in *in vivo* and *in vitro* situations, it is believed that the inducer acts by altering the permeability of the cell membrane to ions and perhaps small substrates.

References

BURNETT, A. L.: The growth process in hydra. J. exp. Zool. 146, 21—84 (1961).
— A model of growth and cellular differentiation in Hydra. Amer. Naturalist 100, 165—189 (1966).
—, L. DAVIS, and F. RUFFING: A histological and ultrastructural study of germinal differentiation of interstitial cells arising from gland cells in *Hydra viridis*. J. Morph. 120, 1—8 (1966).
—, and N. DIEHL: The nervous system of *Hydra*. I. The structure, origin and distribution of nerve elements. J. exp. Zool. 157, 217—226 (1964a).
— — The nervous system of *Hydra*. III. The initiation of sexuality with special reference to the nervous system. J. exp. Zool. 157, 237—250 (1964b).
— —, and F. DIEHL: The nervous system of *Hydra*. II. Control of growth and regeneration by neurosecretory cells. J. exp. Zool. 157, 227—236 (1964).
—, and F. RUFFING: In preparation (1967).
CAMPBELL, R. D.: Tissue dynamics of steady state growth in *Hydra littoralis*. I. Patterns of cell division. Develop. Biol. 15, 487—502 (1966).
CLARKSON, S. G., and L. WOLPERT: Bud morphogenesis in *Hydra*. Nature (Lond.) 214, 780—783 (1967).
DAVIS, L. V.: Inhibition of growth and regeneration in *Hydra* by crowded culture water. Nature (Lond.) 212, 1215—1217 (1966).
—, A. L. BURNETT, J. HAYNES, and V. MUMAW: A histological and ultrastructural study of dedifferentiation and redifferentiation of digestive and gland cells in *Hydra viridis*. Develop. Biol. 14, 307—329 (1966).
HAYNES, J., and A. L. BURNETT: Dedifferentiation and redifferentiation of cells in Hydra viridis. Science 142, 1481—1483 (1963).
LENICQUE, P. M., and M. LUNDBLAD: Promotors and inhibitors of development during regeneration of the hypostome and tentacles of *Clava squamata*. Acta Zool. 47, 185—195 (1966).
LESH, G. E., and A. L. BURNETT: An analysis of the chemical control of polarized form in *Hydra*. J. exp. Zool. 163, 55—77 (1966).
MACKLIN, M., and A. L. BURNETT: Control of differentiation by calcium and sodium ions in *Hydra pseudoligactis*. Exp. Cell Res. 44, 665—668 (1966).
PHILPOTT, D., A. CHAET, and A. L. BURNETT: A study of the secretory granules of the basal disk of Hydra. J. Ultrastruct. Res. 14, 46—36 (1966).
ROSE, S. M., and F. ROSE: The role of a cut surface in *Tubularia* regeneration. Physiol. Zool. 14, 328—343 (1941).
TARDENT, P.: Principles governing the process of regereration in hydroids, p. 21—43. In: Developing cell systems and their control (D. RUDNICK, ed.). 18th Symposium of the Society for the Study of Development and Growth. New York: Ronald Press 1960.

The Multipotential Cell and the Tumor Problem*

ARMIN C. BRAUN

The Rockefeller University, New York

One of the most firmly held beliefs in the field of oncology is that once a cell has been converted into a true tumor cell the resulting heritable change is of a permanent and irreversible type. This concept of irreversibility has for about a century now so completely dominated the thinking of cancer biologists that until quite recently few attempts were made to determine whether it might, in fact, be possible to cause a true tumor cell to revert to a benign or normal state under appropriate experimental conditions. There are now a number of well documented examples ranging from tumors of higher plant species to those of man in which a reversal of the neoplastic state has been achieved experimentally. Several examples of this type will be described and the possible significance of these findings as they relate to the nature of the heritable change will be discussed.

The tumor problem is basically a problem of growth. It is in its very essence a dynamic problem of abnormal and autonomous cell growth and division. A fully autonomous rapidly growing tumor cell may, in fact, be described as a highly efficient proliferating system the energy of which is directed largely toward a synthesis of substances required specifically for continued cell growth and division. The transition from a normal cell to a tumor cell must, therefore, involve a radical and persistent reorientation of biosynthetic metabolism going from the precisely regulated metabolism concerned with differentiated function, which is characteristic of a normal resting cell, to one involving the persistently increased synthesis of nucleic acids, mitotic and enzymatic proteins as well as other substances that are required specifically for continued cell growth and division. It would appear, therefore, that an understanding of the tumor problem would require a characterization at a physiological and biochemical level of the cellular mechanisms that are responsible for this major and persistent switch in biosynthetic metabolism. Since a century of experience has now made it abundantly clear that the transformation of a normal cell to a tumor cell involves a heritable change, an understanding of the tumor problem at another level would appear to require insight into the nature of the heritable cellular change that results in a capacity for autonomous growth of a tumor cell. In considering those matters I should like to use as an experimental model the non-self-limiting neoplastic disease of plants that is commonly known as crown gall. This disease is initiated by a

* The investigations reported here were supported in part by a research grant (PHS grant CA-06346) from the National Cancer Institute, U.S. Public Health Service; a research grant (NSF grant GB-197) from the National Science Foundation; by a Maude K. Irving Memorial Grant for Cancer Research and a Berte Abramson Memorial Grant for Cancer Research (grants E-159B and E-159C) from the American Cancer Society, Inc.

tumor-inducing principle (TIP) that is elaborated by a specific bacterium. The TIP possesses the ability to transform regularly normal plant cells to tumor cells in short periods of time (BRAUN 1943, 1947). Once the cellular transformation has been accomplished, the continued abnormal and autonomous proliferation of the affected host cells becomes an automatic process which is entirely independent of the inciting bacteria or any other readily recognizable infective agent. Fragments of sterile tumor tissue, but not thoroughly ground tumor cells, when implanted into healthy hosts of the same species from which they arose, develop again into crown-gall tumors that are similar in every respect to those initially produced by the bacteria except that the new growths are sterile (BRAUN and WHITE 1943; WHITE and BRAUN 1942). Such tumors, which arise entirely from the implanted cells, are therefore transplantable. Crown-gall tumor cells of many different plant species have been found, moreover, to grow profusely and indefinitely on a simple chemically defined culture medium which consists only of mineral salts, sucrose and three vitamins. This medium does not support the continued growth of normal cells of the type from which the tumor cells were derived. The fact that such tumor cells are transplantable and that they grow profusely and continuously on a simple chemically defined culture medium which does not support the growth of normal cells indicates that a profound and heritable change has occurred as a result of a transformation process. It has been found, moreover, that the transformation of a normal plant cell to a fully autonomous and rapidly growing tumor cell does not occur in a single step but takes place gradually and progressively over a 3 to 4-day period (BRAUN 1943, 1947). By interrupting the transformation process by means of a thermal treatment at any desired time it was possible to obtain stable cultures which possessed varying grades of neoplastic change ranging from very slowly growing benign to rapidly growing fully autonomous tumor cell types. The degree of neoplastic change achieved appears, moreover, to reflect the stage in the normal wound healing cycle in which the cellular transformation occurs. It is just before or during the earliest stages of active cell division in the normal wound healing cycle that normal plant cells are transformed into tumor cells of the most rapidly growing type (BRAUN 1952; BRAUN and MANDLE 1948; LIPETZ 1966).

The question arises, therefore, why tumor cells proliferate continuously on a simple chemically defined culture medium which does not support the growth of normal cells of the type from which the tumor cells were derived. In order to answer this question one must obviously know something about the factors which regulate normal cell growth and division. Growth in plants, as in animals, results either from an enlargement of cells or from the combined processes of cell enlargement and cell division. These fundamental growth processes are controlled in plants by the quantitative interaction of two growth-regulating hormones, the auxins and the cytokinins. The auxins are concerned with cell enlargement while the cytokinins act synergistically with the auxins to promote growth accompanied by cell division (JABLONSKI and SKOOG 1954). These two hormones have been found to play a central role in the development of a capacity for autonomous growth of the plant tumor cell (BRAUN 1956). By selecting three tumors which showed different grades of neoplastic change ranging from very slowly growing benign to rapidly growing fully autonomous cell types and comparing the growth of these, on a simple chemically defined culture medium, with normal cells of the type from which the tumor cells were derived, it was possible to characterize the essential metabolites required for rapid autonomous growth (BRAUN 1958). In those studies the rapidly growing fully autonomous tumor cell was used as the standard. This cell type can synthesize in optimal or nearly optimal amounts all of the growth factors required for its continued rapid growth from mineral salts, sucrose and three vitamins present in the basic culture medium. A moderately fast growing tumor cell type required that the basic medium be supplemented with glutamine, *myo*inositol and the cell enlargement hormone, auxin, to achieve a growth rate comparable to that of the fully transformed cell grown on the basic culture medium. A very slow growing tumor cell type required, in addition to those three substances, exogenous asparagine as well as a purine and particularly a pyrimidine for rapid growth in culture. Those studies demonstrate, then, that as the crown-gall tumor cell becomes more autonomous, its exogenous requirements for rapid growth become less exacting. The normal cells do not grow at all on the basic culture medium. Thus, although the difference between the three types of tumor cells is quantitative since all three can grow continuously but at different rates on the basic culture medium, the difference between the tumor cells and the normal cells is qualitative. The exogenous requirements for the rapid growth of normal cells were found to be the same as those required for the rapid growth of the most slowly growing type of tumor cell, with one exception. The normal cells, unlike the tumor cells, possess an absolute exogenous requirement for a cytokinin, the hormone that promotes cell division. The tumor cells have acquired, as a result of the cellular transformation, the capacity to synthesize this hormone. This, then, represents a fundamental difference between a normal cell and a crown-gall tumor cell since it is the continued production of that newly synthesized metabolite which keeps the tumor cells dividing continuously. This substance, which is a nicotinamide derivative, has now been isolated in pure form (WOOD 1964) and its possible relationship to the 6-substituted purines, which have been known for more than a decade to promote cell division, has been studied (WOOD and BRAUN 1967).

Results of the experiments described above give a great deal of information concerning the workings of the crown-gall system. They indicate that, as a result of the transition from a normal cell to a fully autonomous rapidly growing tumor cell, a series of quite distinct but well defined biosynthetic systems, which represent the entire area of metabolism concerned with cell growth and division, become progressively and persistently activated. On the other hand, the degree of activation of these systems determines the rate at which the crown-gall tumor cell grows.

It is also clear from the results reported above that autonomy, in this instance, finds its explanation in terms of cellular nutrition. The tumor cells have acquired, as a result of their transformation, a capacity to synthesize all of the essential metabolites which their normal counterparts require but cannot make for continued cell growth and division.

Finally, these findings demonstrate that as a result of the transition from a normal cell to a fully autonomous tumor cell, a profound and persistent reorientation in the pattern of synthesis occurs. This switch in metabolism from that found in a normal resting cell to that present in a persistently dividing cell is triggered by irritation accompanying a wound. It may be permanently fixed in the plant tumor cell not only by TIP associated with the crown-gall disease but by other tumorigenic agents as well. This pattern of synthesis is maintained in the plant tumor cell because the two hormones which regulate cell growth and division are continually synthesized by such cell types. The other metabolites shown to be produced in excessive amounts by tumor cells are required for the synthesis of nucleic acids, mitotic and enzymatic proteins and, in the case of inositol, the membrane systems of the cell. Those metabolites are essential to permit the pattern of synthesis concerned with cell growth and division to be expressed.

The question arises how the diverse biosynthetic systems representing the entire area of metabolism concerned with cell growth and division are rendered functional in the crown-gall tumor cell. It is clear that some very fundamental cellular mechanisms must be involved in the simultaneous or perhaps sequential activation of those essential biosynthetic systems. Studies designed to elucidate this aspect of the problem have shown that progressive alterations in the properties of the membrane systems accompany the transition from a normal cell to a fully autonomous tumor cell (BRAUN and WOOD 1962; WOOD and BRAUN 1961). These studies have indicated, moreover, that five and in part six of the seven essential biosynthetic systems shown to be activated in the crown-gall tumor cell are rendered functional by specific ions. Only the system concerned with the synthesis of the cell division promoting hormone, cytokinin, cannot be accounted for on this basis. The observed changes in the properties of the membrane systems appear to reflect progressive increases in the permeability of the tumor cell membranes not only for ions but for certain organic solutes as well (WOOD and BRAUN 1965). The fully autonomous tumor cells take up ions very effectively from dilute salt solutions; the normal cells do not. This, then, represents a fundamental difference between a rapidly growing crown-gall tumor cell and its normal counterpart since a large segment of the area of metabolism concerned specifically with cell growth and division is rendered functional by specific ions.

The essential metabolites (BRAUN and WOOD 1962; WOOD and BRAUN 1961) including the cytokinins (WOOD and BRAUN 1967), which are needed for continued cell growth and division, appear to be the same in normal and tumor cells. This fact suggests that the synthesis of those metabolites by tumor cells results from a derepression of the genome of the host cell. It seems reasonable to assume, therefore, that if new genetic information is introduced into the cells at the time of their transformation, such information acts in some indirect way to derepress that segment of the genome concerned with cell growth and division rather than by supplying directly the information needed for the synthesis of the essential metabolites.

It is a generally accepted belief that heritable changes of the type which we have been discussing are permanent and irreversible. This would certainly appear to be true of the typical crown-gall tumor cell since such cell types isolated from many different plant species have been kept under observation for more than a decade without showing the slightest tendency to become less autonomous.

In studying the origin of the crown-gall tumor cell attempts were made to distinguish between somatic mutation at the nuclear gene level, involving the deletion or permanent rearrangement of genetic information, on the one hand, and epigenetic changes which are concerned merely with alterations in the expression of the genetic potentialities normally present in a cell, on the other. Unfortunately, it is not possible to carry out breeding experiments of the classical type in this system and similar systems and thus determine whether segregation of a possible mutation occurs at meiosis according to the Mendelian laws of heredity. Other less direct methods

9*

had, therefore, to be applied. For studies of this type plants offer distinct advantages as experimental test objects for essentially two reasons. First, certain somatic cells of some plant species remain totipotent throughout the life of a plant. The second advantage is the unique manner in which dicotyledonous plant species grow. Primary growth of such plant species results from the rapid division and subsequent elongation of the meristematic cells at the extreme apex of a shoot or a root. By combining the properties of totipotency and unique growth characteristics it was possible to achieve regularly a recovery — that is, a normalization, of the crown-gall tumor cell and thus gain insight into the nature of the heritable cellular change.

The typical crown-gall tumor cell possesses a pronounced capacity for proliferation, a limited capacity for differentiation and such tumor tissue lacks entirely the ability to organize structures such as roots, leaves or buds. If, however, totipotent cells of certain dicotyledonous plant species are transformed to a moderate degree but not fully, a morphologically very different type of growth results (BRAUN 1953). Such a new growth or teratoma is composed of a chaotic assembly of tissues and organs that show varying grades of morphological development. Sterile tissue fragments isolated from abnormal but organized structures found on the teratomas grow profusely and indefinitely, as do fully transformed tumor cells, on a basic culture medium that does not support the continued growth of normal cells of the type from which the tumor cells were derived. Teratoma tissue differs from typical crown-gall tumor tissue in that it retains indefinitely in culture a pronounced capacity to form abnormal leaves and buds. Such tumors are transplantable. That the teratomas are composed entirely of tumor cells and not of a mixture of normal and tumor cells was demonstrated unequivocally by isolating a number of clones derived from single cells and demonstrating that these clones behaved in every way as did the teratomas from which they were derived (BRAUN 1959).

Since teratoma tissue clones possessed a capacity to form tumor buds, they were admirably suited and were used for studies dealing with the nature of the heritable cellular change that leads to a capacity for autonomous growth of the crown-gall tumor cell. In these studies it was hypothesized that if tumor shoots derived from the abnormal tumor buds found on teratomas could be forced into rapid but organized growth, a recovery from the tumorous state might regularly occur if the primary cellular change leading to autonomy was of an epigenetic type but not if it involved somatic mutation at the nuclear gene level. The results of those studies demonstrated that if tumor shoots derived from tumor buds were forced into rapid but organized growth by means of a series of tip graftings to healthy plants, they gradually recovered from the tumorous state and ultimately became normal in every respect. In view of the results of these studies, somatic mutation at the nuclear gene level seems highly unlikely since the nuclei of normal and tumor cells appear to be genetically equivalent. These findings suggest, rather, that in this instance we are dealing merely with a change in the expression of genetic information present in a cell. The biosynthetic systems concerned with cell growth and division are gradually and progressively activated during the transformation process, while those systems appear again to be gradually repressed during the recovery phase.

The question arises whether a reversal of the neoplastic state is something which is unique to plants or whether it has broader biological implications. Within the past several years there have been a number of well documented instances of recovery from the tumorous state in animals. The reported instances, as in the case of the crown gall disease, have commonly involved multipotential tumor cells.

The first of these, reported by SEILERN-ASPANG and KRATOCHWIL (1962, 1963), deals with a malignant tumor of the European newt, an animal well known for its pronounced regenerative capacities. Following injection of carcinogenic hydrocarbons under the skin of newts, tumors arose in multicentric fashion from the basal cells of the mucous glands of the skin. Such tumors coalesced to form rapidly expanding growths which infiltrated and destroyed normal tissues and metastasized freely to other parts of the animals. Many of the animals died from those complications, indicating the malignant nature of the new growths. In many other instances, however, a most remarkable thing happened. After such tumors had reached large size, infiltrated and metastasized, all of the tumor cells reverted to normal cell types. When such a reversion occurred, the animals recovered completely from the disease. Such a recovery was most readily accomplished when the carcinogen was injected at a site close to the base of the tail and a part of the tail was removed to stimulate the regenerative processes in the animal.

It can, of course, be argued that the newt, like the plant, is a rather low form of life and what happens in the newt or the plant has really very little to do with what may, in fact, occur in higher organisms. What, then, do we know about the occurrence of this phenomenon in mammals? An interesting example of a reversal of the cancerous state as it occurs in the mouse has recently been reported by KLEINSMITH and PIERCE (1964). These workers studied a teratocarcinoma of the mouse. This particular teratocarcinoma was a malignant tumor of testicular origin that contained derivatives of all three embryonic germ layers. Typically, the tumor contains bone, muscle, cartilage, brain, nerve tissue, tooth buds, hair follicles, and many other well differentiated cell types which are present in a completely disorganized array. Interspersed among well differentiated cell types there are undifferentiated cells, the embryonal carcinoma cells, which are the malignant component of the tumor. This broad spectrum of cell types is reproduced indefinitely during serial transfer of the tumor from mouse to mouse. KLEINSMITH and PIERCE were interested in the origin of the well differentiated cell types found in the tumor. Did these differentiated cell types arise and proliferate independently or did all develop from a common precursor or stem cell? In order to test the latter hypothesis, single embryonal carcinoma cells were injected into peritoneal cavities of mice. Of the approximately 50 of these single cells that established themselves and grew, all developed into malignant tumors that killed the animals in a few weeks time. An examination of the tumors showed that they were commonly composed of eight to fourteen different well differentiated cell types together with embryonal carcinoma cells. The differentiated cell types were found to be non-malignant. The results of these studies demonstrate, then, that the embryonal carcinoma cell is a multipotential cell being highly malignant in the undifferentiated state but capable of giving rise to well differentiated cell types which have lost their malignant properties.

Finally, we come to human tumors and the question as to whether regressions involving the maturation and differentiation of tumor cells may occur in man. The best documented examples of this are found in the neuroblastomas. The neuroblastoma is a highly malignant tumor that is derived during organogenesis from the primitive sympathetic nerve cell, the neuroblast. This tumor is commonly diagnosed during the first year of life and may even develop prenatally. In the vast majority of patients with untreated neuroblastomas, metastases soon develop and there is a rapid downhill course culminating in death in a relatively few months. There are striking exceptions to this, however. In 1927 Cushing and Wolbach reported the now classical and much studied case in which all of the cells of a malignant metastasizing neuroblastoma were converted into mature, highly differentiated ganglion cells which had lost their malignant properties. In 1963 Visfeldt described in detail another instance of this type involving the conversion of a neuroblastoma into a benign ganglioneuroma. In addition, Everson and Cole (1966) list 29 well documented cases of the spontaneous regression of neuroblastomas in man.

It has recently been found (Goldstein, Burdman and Journey 1964) that the conversion of neuroblastoma cells into ganglion cells may be accomplished under controlled conditions in culture. Thus, the phenomenon of a reversal of the tumorous state is observed occasionally not only in patients but in cell culture as well.

The examples of recovery cited above represent, of course, exceptional cases. It is, nevertheless, true that it is often through an understanding of such special cases that insight is gained into complex biological phenomena.

Why, then, may multipotential cells recover from the tumorous state while the generality of tumor cells do not? The answer to this question would appear to rest in the fact that multipotential cells are endowed with broad morphogenetic or regenerative capacities. Such cell types can very effectively remodel metabolic patterns, as is clearly evident from the results reported here. The apparent irreversibility of the tumorous state in the great majority of instances may simply reflect an inability of most differentiated somatic cells to undergo intracellular regenerations of the type characteristic of a multipotential cell.

What, then, can be learned from those special cases in which a recovery from the tumorous state has been demonstrated experimentally? It is clear, first of all, that tumorigenesis need not, as is so commonly believed, involve a change in the integrity of the genetic information that is normally present in a cell. It would be difficult, indeed, to accept the explanation that reversions, which as we have seen occur regularly in some tumor systems, represent a controlled series of back mutations.

These systems, on the other hand, provide the best evidence yet available to indicate that the tumorous state generally may be mediated through purely epigenetic mechanisms. The results reported above suggest, therefore, that the tumor problem may be basically a problem of anomalous differentiation and that neoplastic growth, like normal developmental processes, may stem from epigenetic modifications against a constant cellular genome. If this is true, then the cellular mechanisms which normally regulate differentiation should apply to the tumor problem as well. It is unfortunately precisely in this area that present knowledge is inadequate. Nevertheless, such an approach offers the attractive possibility that we may ultimately learn, as the multipotential cell has learned, how to switch the pattern of synthesis in a cell from one that makes it grow as a malignant cell to one that will restore its normal or, in those instances in which extensive deletions have occurred, at least its benign behavior.

References

Braun, A. C.: Studies on tumor inception in the crown-gall disease. Amer. J. Bot. 30, 674—677 (1943).

BRAUN, A. C.: Thermal studies on the factors responsible for tumor initiation in crown gall. Amer. J. Bot. **34**, 234—240 (1947).
— Conditioning of the host cell as a factor in the transformation process in crown gall. Growth **16**, 65—74 (1952).
— Bacterial and host factors concerned in determining tumor morphology in crown gall. Bot. Gaz. **114**, 363—371 (1953).
— The activation of two growth-substance systems accompanying the conversion of normal to tumor cells in crown gall. Cancer Res. **16**, 53—56 (1956).
— A physiological basis for autonomous growth of the crown-gall tumor cell. Proc. nat. Acad. Sci. (Wash.) **44**, 344—349 (1958).
— A demonstration of the recovery of the crown-gall tumor cell with the use of complex tumors of single-cell origin. Proc. nat. Acad. Sci. (Wash.) **45**, 932—938 (1959).
—, and R. J. MANDLE: Studies on the inactivation of the tumor-inducing principle in crown gall. Growth **12**, 255—269 (1948).
—, and P. R. WHITE: Bacteriological sterility of tissues derived from secondary crown-gall tumors. Phytopathology **33**, 85—100 (1943).
—, and H. N. WOOD: On the activation of certain essential biosynthetic systems in cells of *Vinca rosea L.* Proc. nat. Acad. Sci. (Wash.) **48**, 1776—1782 (1962).
CUSHING, H., and S. B. WOLBACH: The transformation of a malignant paravertebral sympathicoblastoma into a benign ganglioneuroma. Amer. J. Path. **3**, 203—216 (1927).
EVERSON, T. C., and W. H. COLE: Spontaneous regression of cancer. Philadelphia-London: W. B. Saunders Co. 1966.
GOLDSTEIN, M. N., J. A. BURDMAN, and L. J. JOURNEY: Long-term tissue culture of neuroblastomas. II. Morphologic evidence for differentiation and maturation. J. nat. Cancer Inst. **32**, 165—199 (1964).
JABLONSKI, J. R., and F. SKOOG: Cell enlargement and cell division in excised tobacco pith tissue. Physiol. Plantarum **7**, 16—24 (1954).
KLEINSMITH, L. J., and G. B. PIERCE JR.: Multipotentiality of single embryonal carcinoma cells. Cancer Res. **24**, 1544—1551 (1964).
LIPETZ, J.: Crown gall tumorigenesis. II. Relations between wound healing and the tumorigenic response. Cancer Res. **26**, 1597—1605 (1966).
SEILERN-ASPANG, F., and K. KRATOCHWIL: Induction and differentiation of an epithelial tumour in the newt (*Triturus cristatus*). J. Embryol. exp. Morph. **10**, 337—356 (1962).
— — Die experimentelle Aktivierung der Differenzierungspotenzen entarteter Zellen. Wien. klin. Wschr. **75**, 337—346 (1963).
VISFELDT, J.: Transformation of sympathicoblastoma into ganglioneuroma. With a case report. Acta path. microbiol. scand. **58**, 414—428 (1963).
WHITE, P. R., and A. C. BRAUN: A cancerous neoplasm of plants. Autonomous bacteria-free crown-gall tissue. Cancer Res. **2**, 597—617 (1942).
WOOD, H. N.: The characterization of naturally occurring kinins from crown gall tumor cells of *Vinca rosea L.* Colloq. intern. centre natl. recherche sci. (Paris), No. 123 (1963), Régulateurs Naturels de la Croissance Végétale, 97—102. Paris: Centre National de la Recherche Scientifique 1964.
—, and A. C. BRAUN: Studies on the regulation of certain essential biosynthetic systems in normal and crown-gall tumor cells. Proc. nat. Acad. Sci. (Wash.) **47**, 1907—1913 (1961).
— — Studies on the net uptake of solutes by normal and crown-gall tumor cells. Proc. nat. Acad. Sci. (Wash.) **54**, 1532—1538 (1965).
— — The role of kinetin (6-furfurylaminopurine) in promoting division in cells of *Vinca rosea L.* Ann. N. Y. Acad. Sci. **144**, 244—250 (1967).

The Stability of the Determined State in Cultures of Imaginal Disks in Drosophila

Walter Gehring*, **

Zoological Institute, University of Zürich, Switzerland

I. Determination and Differentiation

The development of different cell types in higher organisms, arising from one cell type (the zygote), is designated as *cell differentiation*. Differentiation is initiated by the process of *determination* which programs the cells for their future developmental pathway. There is accumulating evidence that the developmental programs reside in the genome and represent sets of genes acting in a coordinate fashion. The fundamental problem of how different genes are coordinately controlled remains to be elucidated. Two major categories of determinative processes are known: one is *embryonic induction* which involves cell or tissue interactions by means of diffusible inducing substances, and another is *cytoplasmic segregation* which is based upon factors intrinsic to the differentiating cells. In the latter case differentiation may be achieved by unequal distribution during cell division of cell components acting as determinative factors. Since determination creates the first essential differences among cells, an understanding of this process might be a key to the problem of differentiation in general. One approach to elucidate determination is to isolate the inducing substances produced by the inducer tissue and to study their action on the reacting tissue. Another approach, which will be considered in this chapter, is to study determination in "mosaic systems" in which intrinsic factors are responsible for setting up the developmental program.

The *imaginal disks* of *Drosophila* are a highly suitable material for such an approach, since a method has been devised for culturing disk cells permanently in a determined state (Hadorn 1963). In addition, Drosophila offers the possibility of an embryological and biochemical as well as a genetic investigation. Hitherto, the emphasis was mainly on the embryological and genetic approach which will be discussed in this survey.

The aim of this chapter is to give a brief review of the experimental results so far obtained and to discuss the conclusions that can be drawn from them. The latter may still be preliminary but they will perhaps stimulate further investigations in the field. The main topic concerns the stability of the determined state and the qualitative changes of determination occuring in cultured cells, since these phenomena might

---

* I am grateful to Professors E. Hadorn and A. Garen for their critical reading of the manuscript.

** Present adress: Department of Molecular Biophysics, Kline Biology Tower, Yale University, New Haven, Connecticut 06520.

give the insight into the problem of determination. Hopefully some of the readers might become "determined" to enter this field, which offers ample possibilities for an experimental approach to this intriguing problem.

II. Imaginal Disks and their Culture *in vivo*

The *Drosophila* larva contains in addition to its larval structures a number of primordia, called *imaginal disks*, which during metamorphosis give rise to specific structures of the adult fly. At the pupal stage most of the larval tissues are histolyzed and the imago (adult) is formed anew from the imaginal primordia (see Bodenstein

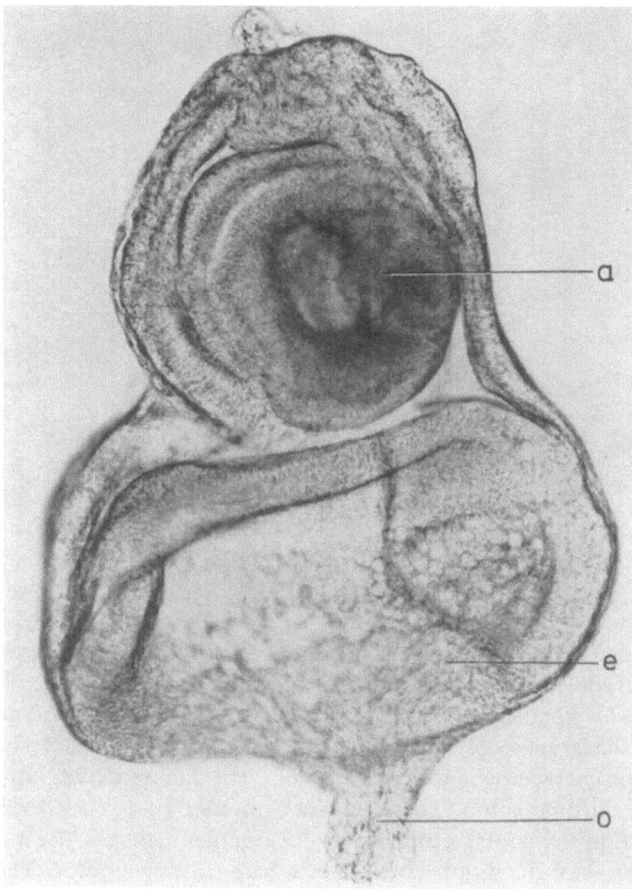

Fig. 1. Eye-antennal disk of a mature larva. Living unstained preparation. a = antennal disk, e = eye disk, o = optic stalk. (Magnification 80 ×)

1950). There are three pairs of disks forming the *head* of the fly: the labial disks, the imaginal cells of the clypeo-labrum (Gehring and Seippel 1967) and the eye-antennal disks (Fig. 1). The adult *thorax* and its appendages arise from three pairs of leg disks and three pairs of dorsal thoracic disks forming the dorsal part of the thorax with the wings and the halteres. Finally, the integument of the *abdomen* is derived from small groups of imaginal cells called "histoblasts", and the genital apparatus located in the last abdominal segments is formed by the unpaired genital disk. Separate primordia have been identified for several internal organs like the gut, the salivary glands and the gonads.

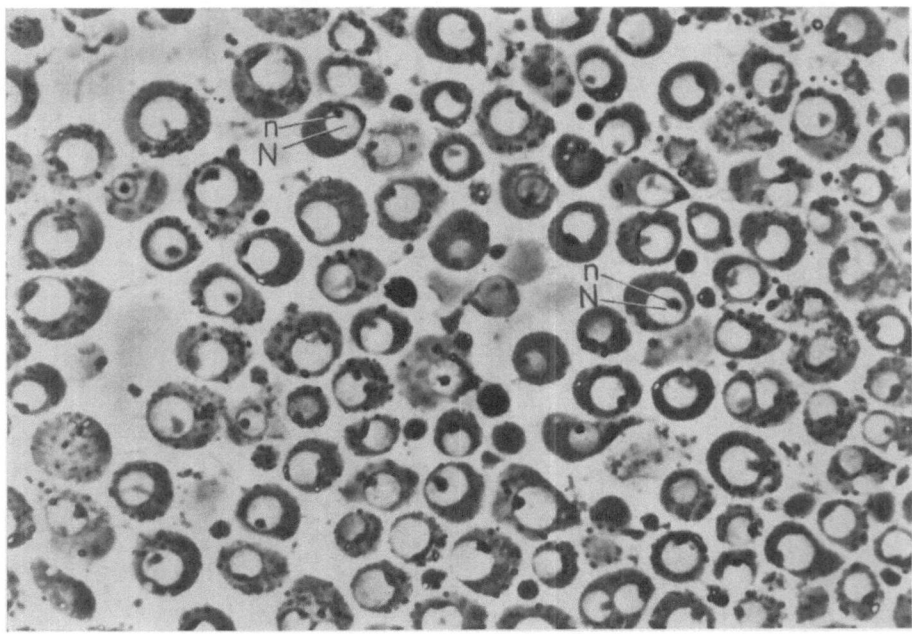

Fig. 2. Living cells from a dissociated eye-antennal disk. Phase contrast. N = nucleus, n = nucleolus. (Magnification 800 ×). By courtesy of P. SUTTER, P. FELS, and H. FRISCH-KNECHT

The imaginal disks first appear as invaginations or thickenings of the epidermis in the later embryonic stages. It is not known whether all the disk cells are of ectodermal origin or whether there is a contribution from invading mesodermal cells as well. During the larval stages the disks grow considerably by cell division, acquiring a definite shape characteristic for each kind of disk; differentiation into adult structures is delayed until pupation. The disks are composed of epithelial cell layers which become folded in the course of the larval development and enclose a narrow lumen. The outer surface of the disk is covered by a basement membrane. The disk cells are small basophilic cells about 5 μ in diameter, with a relatively large nucleus and little surrounding cytoplasm. In Fig. 2 living cells of a dissociated eye-antennal disk are shown in phase contrast. The cells appear fairly uniform, but an analysis of their fine-structure which might reveal differences among the cells has not yet been accomplished. WADDINGTON and PERRY (1960) have described clusters of cells in the eye disk which appear to represent the anlagen of the ommatidia and can be distinguished cytologically from the remaining disk cells.

Disks can be transplanted into larval or adult hosts by a method devised by EPHRUSSI and BEADLE (1936) (see also URSPRUNG 1967). The donor larva is dissected and the isolated disk is injected by means of a micropipette into the body cavity of a host animal. When a disk is transplanted into a *larval host* of the same age as the donor, it develops synchronously with the host and undergoes metamorphosis. After the host fly has hatched, the implant can be removed from its abdomen and the structures formed by the implant examined.

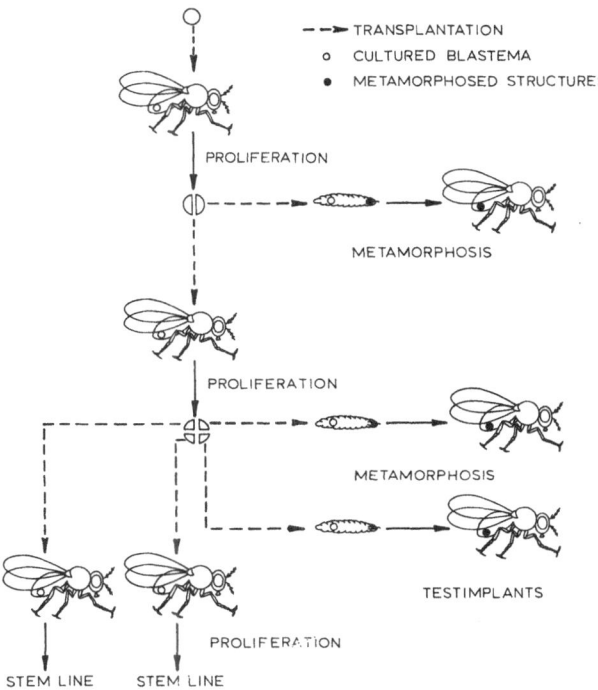

Fig. 3. Method for culturing imaginal disks *in vivo* (modified after HADORN 1963). Explanation in the text

Using the method of HADORN (1963) blastemas of imaginal disks can be cultured indefinitely in the abdomen of adult flies where they do not undergo metamorphosis. The host's hemolymph serves as a culture medium which allows proliferation, but due to a lack of metamorphosis hormones adult differentiation is not induced. Thus, the implant maintains its larval state. Since the culture period is limited by the lifespan of the host fly, the implant must be transferred to a young adult host every 2 to 4 weeks (Fig. 3). At each transfer the implant, which has increased in size, is cut into a number of fragments. Some of these fragments are injected into fresh adult hosts, whereas others are transplanted back into larvae with which they undergo metamorphosis. These larval "test implants" provide information about the capacities of the cultured disk cells for differentiation. It has been found that the cultured cells maintain their capacity for normal differentiation even after several years of culturing (HADORN 1966) and also show the normal karyotype (GEHRING 1966a; REMENSBERGER 1968). Occasionally, loss of the capacity to differentiate and abnormal karyotypes are observed. It should be emphasized that we are dealing with cultures of blastemas rather than cultures of separate cells. However, there are techniques available to study the capacities of single cells in the cultures as well. Isolated cells obtained by dissociation of disks carrying different marker genes can be intermixed, reaggregated, and tested for their developmental behavior (HADORN, ANDERS, and URSPRUNG 1959; NÖTHIGER 1964; GARCIA-BELLIDO 1966a). Another method makes use of somatic crossing-over or somatic mutation to induce genetically marked single cells which will form clones during proliferation in culture (GEHRING 1967). Thus, it is possible to analyze determination at the cellular level, using *in vivo* techniques.

III. Criteria for the Determined State of Imaginal Disk Cells

Several experimental procedures have been used to identify embryonic and larval cells which are determined for differentiation into adult structures, mainly those of the integument in *Drosophila*. Determination already occurs during embryogenesis. Treatment of wild-type embryos with ether or a heat shock produces flies in which the metathorax with its halteres is transformed almost completely into a second meso-thorax with wings (HENKE and MAAS 1946; GLOOR 1947). These ether treated flies show the same phenotype as some *bithorax* (BRIDGES and MORGAN 1923) and *tetraptera* (ASTAUROFF 1929) mutants. The highest frequency of transformed flies was obtained when the ether "pulse" was applied to embryos of the blastoderm stage about 1–3 hours after fertilization. This suggests that determination for differentiation into adult structures occurs early during embryogenesis. By means of UV irradiation of later embryonic stages GEIGY (1931) was able to induce specific defects in certain structures of the adult fly, whereas the larval structures remained unaffected. Flies were obtained, for example, with one leg or wing missing, or with duplications of leg segments. Since in these cases the larval organs were normal, we have to assume that the primordia for the adult structures are already separated from the larval tissues during embryogenesis, apparently before the imaginal disks can be morphologically detected. Furthermore, the elimination of complete sets of organs derived from one disk shows that *the presumptive disks are already programmed to form a fixed area of the adult integument*.

The same conclusion was drawn from cell lineage studies by means of genetically marked cells initiated by STURTEVANT (1929) (see STERN 1963). Genetically marked single cells are produced by somatic mutation or somatic crossing-over. In the course of proliferation the marked cell gives rise to a clone which can be detected in the adult as a patch of marked tissue. When the somatic mutation occurs at a very early embryonic stage, a large fraction of the adult integument consists of marked cells. During later embryonic (or early larval) stages the clones of marked cells become restricted to those areas which are derived from a specific imaginal disk and do not include areas originating from different disks. From these observations and the UV elimination experiments we conclude that during embryogenesis the cells of an imaginal disk become determined to form a specific area of the adult integument which is derived from that disk. At this stage the disk cells still can give rise to all the cell types occurring in this area.

Later in development the number of cell types to which a disk cell can give rise is further restricted. For example, certain cells of the eye disk of a late second instar larva still have two possible pathways of differentiation; they may either differentiate into ommatidia or into cells forming the border of the eye (BECKER 1957). Similarly, certain cells of the male genital disk in the second or even early third larval instar may still give rise to both clasper and lateral plate cells (E. ULRICH and R. NÖTHIGER, unpubl.). Towards the end of the third larval instar the cells become further determined for one of the alternative pathways. At this stage the method of following the cell lineage becomes unapplicable, since the marked cells divide only a few times before metamorphosis and therefore produce only a tiny patch of marked tissue.

During the last (third) larval instar the disks are most accessible to embryological surgery. By cutting the disk from a mature larva into definite fragments which are then transplanted separately into host larvae of the same age, it can be shown that different regions of the disk form different organs. The anlagen for the different organs can in this way be mapped. The most detailed maps have been established for the male and female genital disk (HADORN, BERTANI, and GALLERA 1949; URSPRUNG 1959), the wing disk (HADORN and BUCK 1962) and the male foreleg disk (SCHUBIGER 1968). Another method consists in UV microbeam treatment to eliminate small limited areas of the disks for a more refined analysis (URSPRUNG 1957 and 1959).

Even within an individual organ anlage, further details can be mapped. For example, a small sense organ in the third antennal segment, the sacculus, is always derived from the anterior half of the antennal disk which contains only a part of the anlage for the third antennal segment (GEHRING 1966a). In the trochanter anlage of the male foreleg disk, a region determined for forming a single bristle has been mapped (NÖTHIGER and SCHUBIGER 1966; SCHUBIGER 1968). These results indicate that determination of the imaginal disks occurs in a stepwise manner, progressively restricting the differentiation potential of the cells.

Although the fragmentation experiments prove the mosaic nature of the disk, they give no information about the *state of determination of an individual cell*. To study this problem, isolated cells obtained by trypsinization or mechanical dissociation of disks carrying different marker genes were intermixed, reaggregated and tested in a metamorphosing host for their differentiative capacities (HADORN, ANDERS, and URSPRUNG 1959; URSPRUNG and HADORN 1962; NÖTHIGER 1964; TOBLER 1966; GARCIA-BELLIDO 1966a and 1966b). The reaggregates form organized adult structures which may be composed of cells of one genotype only, or cells of both genotypes forming a genetically mosaic structure. *Drosophila* offers a number of convenient marker genes: *ebony* and *yellow* mutants which produce dark and yellow color, respectively, of the chitinous structures, the *singed* mutant which is characterized by altered bristle shape; the *multiple wing hairs* mutant in which the single cell hairs (trichomes) are replaced by groups of trichomes. Since both bristle shafts and trichomes are each produced by single cells, the differentiation of a single cell can be traced in this way.

In dissociation-reaggregation experiments, mosaic structures composed of cells with different genotypes are formed only by *isotypic* cells (NÖTHIGER 1964). For example, anal plate cells will frequently form mosaics with anal plate cells of a different genotype or sex, but fail to associate with clasper cells. *Heterotypic* cells tend to separate from each other. From this result we conclude that the presumptive anal plate cells differ specifically from the clasper cells. The association of isotypic cells and the separation of heterotypic cells is most likely achieved by cell migration and selective adhesion of the cells, as observed in vertebrate cells, since there is no indication of selective cell death.

If cells from different disks having no isotypic cells in common are intermixed, the cells sort out completely and do not form mosaic structures. The only exceptions to that rule are the so-called "faulty" mosaics which occur very rarely (NÖTHIGER 1964). If, for example, a single cell of a genital disk (or a very small group of cells) is trapped in a large area of wing disk cells, it differentiates autonomously and forms a specific genital bristle. Thus, the surrounding blastema does not exert a determinative influence upon the isolated cell. This shows that the quality (specificity) of determination at this stage is carried by the individual cell rather than by the blastema as a whole.

Are the cells of a disk from a mature larva fully programmed or does further determination occur during the pupal stage? There is some evidence for further determination at the early pupal stage during which cell division continues. In the case of the bristle which consists of two external cells — the shaft cell (trichogen) and the socket cell (tormogen) — the external cells are derived from a bristle mother cell by a differentiative division at this late stage (LEES and WADDINGTON 1942).

Therefore, we are inclined to assume a prefinal determination which leaves the cells a limited array of possible pathways. Furthermore, there is also an indication of cell interactions leading to the determination of a specific cell type after pupation (TOBLER 1966). Certain leg and wing bristles are accompanied by a special trichome called a bract (HANNAH-ALAVA 1958; PEYER and HADORN 1965), occupying a definite position in relation to the bristle. The cell which forms a bract is not derived from the bristle mother cell, since genetic mosaics are observed in which the bract cell differs in its genotype from the bristle cell. Isolated bracts without a corresponding bristle are never observed after reaggregation of isolated cells from mature larvae, which may indicate that the bristle induces the formation of a bract in a neighbouring cell after pupation (TOBLER 1966). However, one might alternatively assume that the presumptive bract cells are determined prior to the dissociation of the disk but do not form a bract if the bristle cell is removed.

In summary the present evidence indicates that the individual cells of an imaginal disk of a mature larva are determined to form one or at most very few specific cell types.

IV. Regulative Versus Mosaic Development

The mosaic nature of the imaginal disks was clearly demonstrated in the fragmentation experiments mentioned above. In those experiments fragments of disks from mature donors were implanted into larvae of the same age, which forces the implant to undergo metamorphosis immediately. The fragment then forms just those structures for which it is determined (mosaic development).

Now we would like to know if, when metamorphosis is delayed to allow further proliferation, a fragment can form a complete set of structures which are normally derived from the entire disk. Metamorphosis can be delayed by transplanting the fragment into a younger host larva or by a two-step experiment, in which the fragment is first cultured in an adult host and then transplanted back into a larva. In both cases the implant proliferates further before differentiation even though it is derived from a disk of a mature larva. The interpretation of the results of such experiments was obscured for a long time by the fact that they were performed on the genital disk which is bilaterally symmetrical; furthermore the complete separation of the different organ anlagen of the genital disk is very difficult. By using an asymmetrical paired disk, such as the wing disk, the following results were obtained which are easier to interpret and which apply to the genital disk as well. According to the anlage map of the wing disk (HADORN and BUCK 1962) the proximal portion of the disk contains the anlagen for the dorsal mesothorax, whereas the distal fragment forms the wing. Proximal fragments which are determined for thorax structures were tested for their ability to "regenerate" wing structures after a prolonged period of proliferation (GEHRING 1966a). All the fragments analyzed showed additional thorax structures, but in no case was regeneration of wing structures observed. This leads to the conclusion that the quality of determination is reproduced by the anlagen during proliferation.

Similar results were obtained with fragments of antennal disks. If an anterior fragment of a mature antennal disk, which is determined to form one palpus, is transplanted into a younger host larva it forms two symmetrical palps of normal size and does not regenerate structures derived from the posterior region of the disk (GEHRING 1966a). These findings have been confirmed for the leg disk by fragmentation experiments and also for the genital disk by the sophisticated method of using UV microbeam treatment to eliminate certain organ anlagen of the disk (NÖTHIGER and SCHUBIGER 1966). We therefore conclude that the fragments of disks show a mosaic development even after prolonged proliferation and give rise only to those structures for which they are determined.

However, the resulting structures are not just oversized organs but represent bilaterally symmetrical duplications of normal sized organs (GEHRING 1966a; WILDERMUTH 1968). This implies a capacity of the disks for regulating size and symmetry of the organs. Since this mechanism involves proliferation, the term *proliferative regulation* (NÖTHIGER and SCHUBIGER 1966) has been proposed. Its mechanism, however, remains to be elucidated.

V. Cell Heredity and Transdetermination

By serial transplantation of imaginal blastemas into adult hosts (see p. 139) the phase of cell proliferation which is characteristic of the larval stage can be prolonged indefinitely. Samples which are transplanted back into larvae after each passage in an adult host (called a transfer generation) provide information about the capacity of the culture for differentiation. Since the original blastema is determined, the question arises whether the quality of determination is replicated in the cultures. In long-term cultures in adult hosts a continuous replication of the original determination qualities is in fact observed. In certain sublines of cultures of male genital disks, cells determined for anal plate formation were continuously reproduced during more than 70 transfer generations over a period of several years (HADORN 1967). It was found that the determination quality is continuously transmitted from one transfer generation to the next. Since the individual cell rather than the blastema as a whole carries the quality of determination (p. 142) we designate this process as *cell heredity*. In certain sublines a loss of the capacity for anal plate formation is observed. This is due to the fact that the cultured blastema is a mosaic of organ anlagen. As the cultured blastema is cut at random into a number of fragments at each transfer to a new adult host, some of the fragments may not contain any cells determined for anal plate

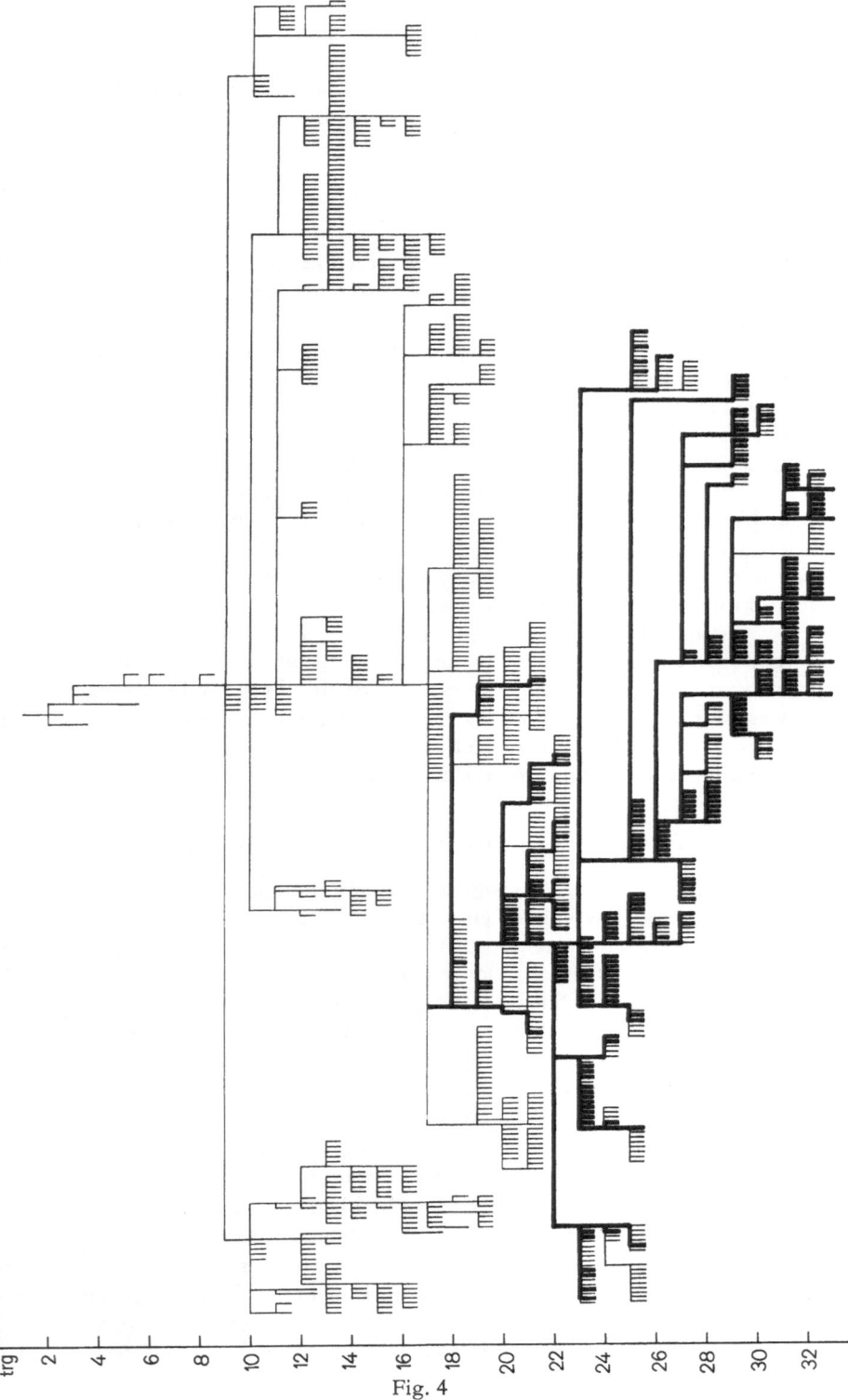

Fig. 4

formation and therefore give rise to sublines which have lost the capacity to form anal plates.

However, in the cultured blastemas occasional *changes in cell heredity* affecting determination occur, which were first observed by SCHLÄPFER (1963) and HADORN (1963). For example, cultured genital disks can give rise not only to genital structures *(autotypic elements)* but also to antennal and leg structures which in normal development are formed by other disks *(allotypic elements)*. The process leading to such allotypic structures is designated as *transdetermination* (HADORN 1965). Between the autotypic and the allotypic structures, a sharp borderline can be recognized; there is no transition zone in which the cells would show some sort of intermediate differentiation with characteristics from both autotypic and allotypic structures. Thus, the different developmental pathways appear to be mutually exclusive. This phenomenon has been designated as "canalization of development" (WADDINGTON 1956).

By means of marker genes it can be shown that the allotypic cells are derived from the implant and not from invading host cells (GEHRING 1966a). When antennal disks carrying the marker gene *yellow* are cultured in hosts homozygous for *ebony* and *multiple wing hairs*, the allotypic structures always show the donor genotype. Furthermore, in cultures of male genital disks grown in female hosts only, allotypic tarsal structures with male sexcombs were observed and all the cultured cells showed a male karyotype (HADORN 1966; P. REMENSBERGER 1968). Thus, a contribution from host cells can be ruled out. However, the possibility could not be excluded that the allotypic cells are derived from undetermined cells contained in the blastema rather than from determined autotypic cells which underwent transdetermination. This possibility will be discussed in the following section.

The information for allotypic differentiation is inherited in the same way as the information for autotypic differentiation. In a "pedigree" of the test implants (p. 139) the continuity of cell heredity is most obvious in those allotypic structures which are rarely formed by transdetermination. An example is illustrated in Fig. 4, which represents the pedigree of the test implants derived from a culture of haltere disks (GEHRING, MINDEK, and HADORN, in preparation). The autotypic haltere elements have a slow rate of proliferation and were therefore lost after 12 transfer generations. The first transdetermination led from haltere-forming cells to presumptive wing cells, which were followed by other allotypic structures not indicated in the pedigree. In the 18th transfer generation, the wing-forming cells gave rise to presumptive eye-cells (dark lines in Fig. 4). This transdetermination step occurs very rarely. After the initial transdetermination the determination for eye-formation is continuously inherited over more than 14 transfer generations and thousands of ommatidia are formed.

VI. Clonal Analysis

The problem whether the allotypic cells are derived from undetermined cells or from determined autotypic cells can be resolved by an analysis of clones derived from single disk cells. However, at the present time we are not able to culture single cells

---

Fig. 4. Cell heredity of the eye-forming cells in a long-term culture initiated by haltere blastemas. "Pedigree" of the testimplants derived from the culture during 32 transfer generations (trg.). Each testimplant is represented by a short vertical line. In the 18th generation transdetermination led to eye-forming cells (dark lines) which reproduced their quality of determination over more than 14 transfer generations

which would form a differentiating clone. This difficulty was overcome by the application of genetic methods (GEHRING 1967). The procedure is illustrated in Fig. 5. Larvae heterozygous for several marker genes were treated with X-rays in order to induce genetically marked single cells in the imaginal disks by somatic crossing-over or by somatic mutation giving rise to homozygous or hemizygous cells (cf. STERN 1936; BECKER 1957). The antennal disks from the irradiated larvae were cultured in adult

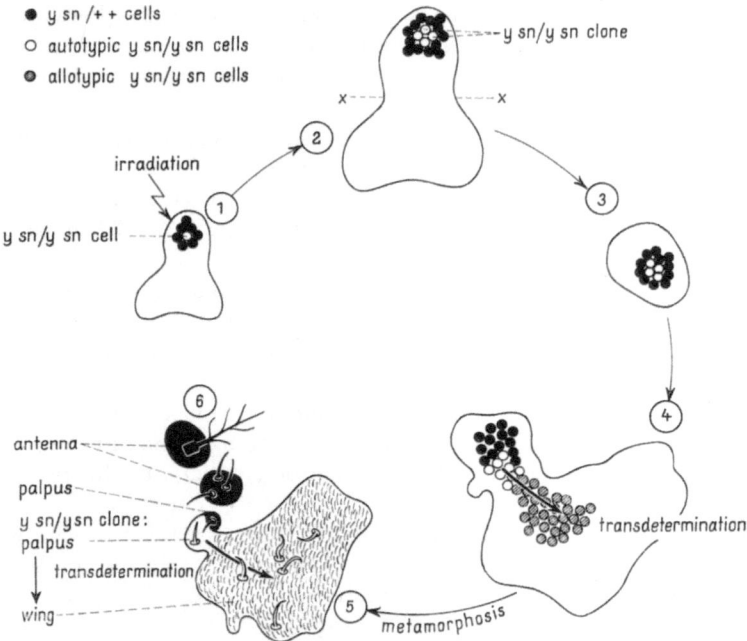

Fig. 5. Induction of clones derived from a genetically marked single cell in cultures of antennal disks (after GEHRING 1967). 1. X-ray treatment of second instar larvae heterozygous for the recessive marker genes *yellow* (*y*) and *singed* (*sn*) and induction of a *y sn*-marked cell (*y sn/y sn*) by somatic crossing-over in the eye-antennal disk. 2. Dissection of the antennal disk from the mature larva. 3. Transplantation into an adult host. 4. Proliferation of the cultured disk and formation of a clone of *y sn/y sn* cells in which transdetermination occurs. 5. Transplantation into a larval host with which the cultured disk undergoes metamorphosis. 6. Metamorphosed structures including heterozygous autotypic cells (*y sn/+ +*) and a clone of *y sn/y sn* cells. The clone consists of autotypic palpus cells and allotypic wing cells. (This particular case is illustrated in Fig. 7 in detail)

hosts, where the marked single cells give rise to a *clone*. The clones can be detected after transplantation back into a larval host. The frequency of somatic crossing-over is sufficiently low that the probability of two independent crossing-overs in one organ anlage is very small. According to our calculations the observed marked areas are therefore derived from a single cell. From a total of 94 marked clones, 78 contained autotypic cells only. The remaining 16 clones listed in Fig. 6 mostly contained both autotypic and allotypic cells. A representative clone of *yellow singed* (*y sn*) marked cells is illustrated in Fig. 7. The autotypic antennal parts are not marked and show wild-type coloration (+). The palpus, which is also an autotypic element formed by the antennal disk, is a genetic mosaic of + and *y sn* areas with twisted yellow bristles. The *y sn* marked base of the palpus is in direct connection with a large allotypic wing area which shows the same phenotype *(y sn)*. Thus, the clone includes part of the palpus region and the whole allotypic wing

Clone number	Autotypic Elements				Allotypic Elements				
	1st Ant. seg.	2nd Ant seg.	3rd Ant. seg.	Palpus	Wing	Head	Tarsus	Coxa	?allo-typic
1				X	X				
2				X	X				
3		X	X		X				
4				X	X	X			
5			X	X	X	X			
6					X	X			
7				X	X				
8				X	X				
9	X					X			
10					X				
11			X				X		
12	X	X		X				X	
13					X				
14					X				
15		X					X		
16									X

Fig. 6. Composition of the clones containing allotypic cells in cultures of antennal disks (after GEHRING 1967). cross-hatched bars: structures contained in the clone. Ant. seg. = antennal segment

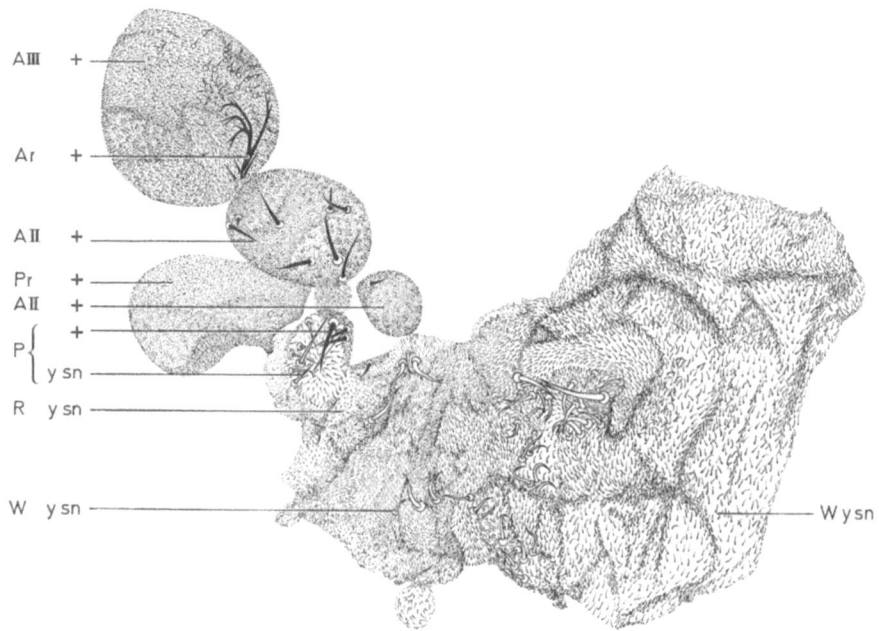

AIII +
Ar +
AII +
Pr +
AII +
P { +
    y sn
R   y sn
W   y sn   W y sn

Fig. 7. Transdetermination in a clone of *y sn*-marked cells derived from a culture of antennal disks (after GEHRING 1967). The clone includes part of the autotypic palpus (P *y sn*) and the rostral membrane (R *y sn*) which gave rise to the allotypic wing (W *y sn*) by transdetermination. The remaining autotypic parts designated with + show wild-type coloration. A II, A III = second and third antennal segment, Pr = prefrons (Magnification 50 ×)

area. This pattern can only be explained by assuming that a somatic crossing-over or mutation occurred in an autotypic cell of the palpus anlage which subsequently gave rise to the other autotypic and allotypic cells of the clone. In five clones a similar pattern of palpus and wing cells was observed. In two clones the autotypic palpus cell had multiplied considerably and formed several palps. This indicates that the clone mother cell was indeed determined for palpus since it transmitted its quality of determination to its descendents over many cell generations. Therefore, we conclude that *transdetermination can occur in clones derived from a determined autotypic cell*. It must be emphasized that although these clones originate from a single autotypic cell, the transdetermination might not have occurred in a single cell of the clone but could have involved a simultaneous response of several cells.

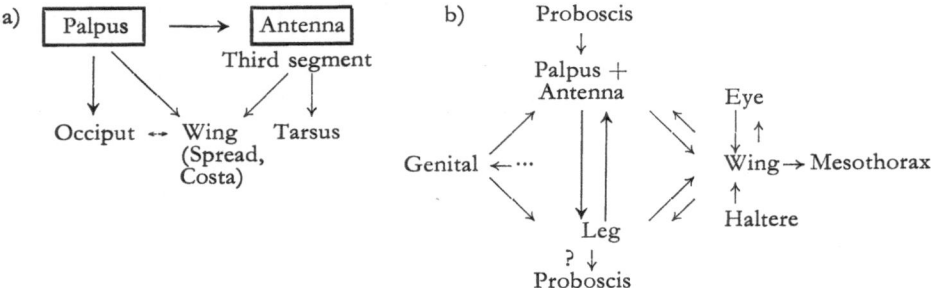

Fig. 8. Schemes of the transdetermination sequences. a) Scheme based on clonal analysis. ▭ autotypic elements. b) Comprehensive scheme summarizing the observations of several authors. Short arrows indicate rare transdeterminations. The dotted arrow indicates that transdetermination to genital cells was observed but it is not known from which cell type these are derived

Transdetermination is a directed process leading from one cell type to a limited array of other cell types. For example, in the clone discussed above, palpus cells gave rise to specific wing cells. The observed *sequences of transdetermination* in the clones listed in Fig. 6 are summarized in the scheme of Fig. 8a. The step from palpus to antenna involves two kinds of autotypic cells both derived from the antennal disk and is therefore designated as *region specific transdetermination* (GEHRING 1966a). A more comprehensive scheme based upon the work of several authors on various disks (see HADORN 1966) is given in Fig. 8b. This scheme is formulated in more general terms, since the evidence for the various steps is rather indirect. For example, haltere stands for all the different cell types constituting the haltere disk, since we do not know yet which cell type undergoes transdetermination to wing-forming cells. As the wing cells arise in cultures of haltere disks before any other allotypic cells are observed (GEHRING, MINDEK, and HADORN, in press), we conclude that the wing cells are derived directly from haltere cells.

The scheme in Fig. 8b shows that certain transdetermination steps are *reversible* as was first shown for the change from the third antennal segment to tarsus (GEHRING 1966a, and SCHUBIGER 1968). Other steps seem to be irreversible. For example, no transdetermination leading to haltere structures has been observed. Thus, certain transdeterminations occur relatively frequently whereas others occur very rarely if ever. The scheme in Fig. 8 apparently does not show any correlation with the arrangement of the various organs *in situ*. The only case indicating such a correlation is the transdetermination from female genital disks to antennal segments, which seem to arise in the same proximo-distal order in which they are arranged *in situ* (MINDEK, in preparation).

The *relative frequency* of transdetermination in repeated experiments remains costant (see Table 1). The only significant difference between the two experiments listed,

Table 1. *Relative frequencies of the different allotypic elements in short-term cultures of antennal disks in independent experiments. The relative frequencies are expressed with respect to the total number (N) of cultures containing allotypic elements*

Allotypic element	Relative frequency	
	1st experiment (N = 54)	2nd experiment (N = 118)
Wing	0.87	0.81
Thorax	0.13	0.17
Tarsus	0.24	0.49*
Proximal leg segments	0.06	0.09
Head	0.44	0.58
Cibarium	0	0.06

* The increased frequency of tarsal structures is due to irradiation in the second experiment.

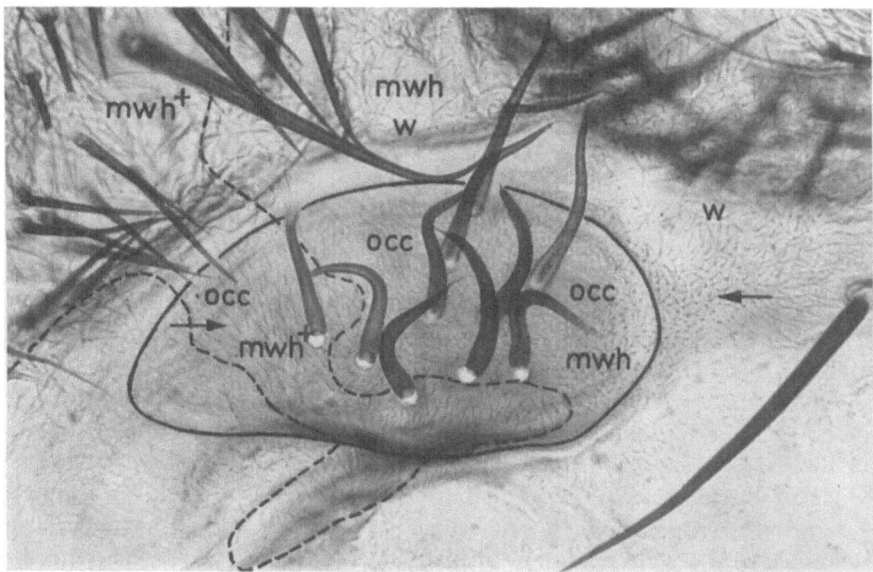

Fig. 9. Clone of genetically marked *(mwh)* cells extending over the border between head (occiput) and wing structures (after GEHRING 1967). In the mutant *multiple wing hairs (mwh)* the cells form groups of shorter trichomes (cell hairs) in a disorderly arrangement whereas the wild-type cells *(mwh+)* have larger single trichomes pointing in the same direction (arrows). The *mwh*-marked cells form a continuous area across the border between the occiput (occ) and the wing (w). This indicates that transdetermination from presumptive occiput to wing cells (or *vice versa*) takes place synchronously in at least two neighbouring cells, a *mwh*-marked and a wild-type cell. ——— border line between head (occiput) and wing structures; -------- border line between *mwh* and wild-type (+) cells

is due to X-irradiation in the second experiment which induced a higher frequency of tarsal structures (see p. 151). Thus, each transdetermination from one cell type to another occurs with a definite probability.

An important problem which can be solved using genetic mosaics is whether a transdetermination occurs independently in individual cells or simultaneously in several neighbouring cells. In our cloning experiments we found several cases in which the allotypic areas were genetic mosaics. However, this finding cannot be considered as evidence for a simultaneous transdetermination in both a marked and a wild-type cell, since isotypic cells tend to associate (p. 141) and the mosaic therefore might be formed by two independent transdeterminations followed by association of the allotypic cells. Nevertheless, if the border line between the areas of marked and wild-type cells in the contact zone between two different organs is carefully analyzed, it is obvious that the distribution of the cells is non-random. An example is illustrated in Fig. 9 which shows the contact zone between two different allotypic structures, an occiput and a wing area. In this case the cells are marked with the gene *multiple wing hairs (mwh)* which leads to the formation of a group of smaller trichomes (cell hairs) instead of a single larger one. The *mwh* cells form a continuous area across the border between the occiput and the wing area. This pattern cannot be explained by random association of cells arising from two independent transdeterminations. Therefore, the same kind of *transdetermination* must have occurred *synchronously* in at least two neighbouring cells, an *mwh* and a wild-type cell.

VII. Possible Causes of Transdetermination

At first thought the *culture medium*, i.e. the adult hemolymph, might be a possible cause of transdetermination. Yet, this possibility is very unlikely since transdetermination occurs both in larval and adult hosts (Hadorn 1963; Gehring 1966a). It was also shown that the frequency of transdetermination does not depend directly on the period of time the blastema is cultured (Tobler 1966). The only factor so far detected which has an influence on the frequency of transdetermination is proliferation. First, a highly positive correlation between the frequency of transdetermination and the rate of proliferation was observed (Tobler 1966). However, a correlation does not necessarily imply a causal relationship. There is another observation of importance in this connection. If a labial disk is cultured *in vivo* it usually duplicates and forms a complete symmetrical proboscis (Wildermuth and Hadorn 1965; Wildermuth 1967). By labelling with thymidine and autoradiography it was shown that replication takes place all over the disk, but the rate of cell division seems to be much lower on the original side of the disk than on the opposite side formed by duplication (Wildermuth, 1968). If transdetermination occurs, the allotypic cells are always found on the "newly formed" side which underwent more replication cycles but is otherwise identical. This observation demonstrates that *cell divisions* are a *necessary prerequisite* for transdetermination. The actual cause of transdetermination is still unknown. It seems to reside in the mechanism of cell heredity itself, which sometimes fails to replicate the quality of determination and thereby causes a transdetermination. A hypothetical scheme for intrinsic cellular mechanisms which might lead to transdetermination was proposed by Hadorn (1967). It is mainly based upon the observation that cell division is a necessary prerequisite for transdetermination. A higher rate of cell division presumably causes a dilution of the carriers of determination which might lead to the activation of other sets of genes by a feedback system. Dilution of the carriers alone would not explain the fact that certain transdeterminations are reversible. The alternative differentiation into mutually exclusive pathways might be explained by assuming one or a very small number of controlling genes regulating integrated sets of genes responsible for the differentiation of an organ (cf. GEHRING 1966a).

Developmental changes leading to the same effect as transdetermination can be produced by mutations, the so-called *homeotic mutations* which have been known for a long time (for discussion of the genetic analysis see LEWIS 1964). These mutations may interfere drastically with determination. Thus, in the recently discovered mutant *Nasobemia* which appears to be a single gene mutant, the antenna and the adjacent part of the head capsule are transformed into a complete middle leg and the corresponding part of the ventral thorax, which normally are derived both from the same leg disk (GEHRING 1966b). This shows that a single gene mutation can bring into action all the genes necessary for the differentiation of a leg disk in a blastema which would otherwise form head structures. The mutant *aristapedia* (Balkaschina 1929) has a more limited effect in that it transforms only the arista and the adjacent part of the third antennal segment into a tarsus. When *aristapedia* antennal disks are cultured and tested in wild-type hosts, tarsal structures are found (GEHRING 1966a). However, after a few transfer generations aristae are also found although this particular mutant (ss^a) never forms an arista *in situ*. This indicates that the mutant has not lost the genetic information to produce an arista. Probably the mutation affects a mechanism controlling which of the two alternate determination programs will be expressed.

From several of these mutants *phenocopies* may be obtained by various treatments. For example, phenocopies of the mutant *aristapedia* are produced by treating the larvae with nitrogen mustard (BODENSTEIN and ABDEL-MALEK 1949), sodium metaborate (SANG and McDONALD 1954), fluoro-uracil or X-rays (GEHRING 1964 and 1967). All these agents seem to have one feature in common, they interfere with growth; but their precise mode of action on the antennal disk is not known. The frequency of phenocopies obtained varies greatly among different strains and therefore depends to a great extent on the genetic constitution and the possible presence of subthreshold mutants. Certain strains do not respond to the treatment at all, which indicates that these phenocopies do not correspond exactly to transdeterminations, since these have been observed in all the strains investigated so far.

As changes of determination are produced by single gene mutations one might assume that transdetermination is due to *somatic mutation*. However, several arguments point against this hypothesis. Although exact data on the absolute frequency of transdetermination for the individual cell are not available, it is evident that this frequency is to be significantly higher than the frequency of spontaneous somatic mutations of marker genes which is very low in the cultures. Furthermore, transdetermination is a directed process leading from one cell type to another with a definite probability and certain cell types revert with a high frequency. Finally, the strongest argument comes from the observation that neighbouring cells can undergo the same kind of transdetermination simultaneously. Thus, it seems unlikely that transdetermination is caused by somatic mutations. But the homeotic mutants show that the inactivation (or activation) of a single gene yields the same result as transdetermination. It seems

probable that the homeotic mutations act in much the same way as transdeterminations and therefore an understanding of one process might give an explanation for
the other.

VIII. Conclusions

In vivo cultures of imaginal disks provide an excellent system to maintain proliferating cells in a determined (programmed) state so that they do not undergo adult
differentiation. Thus, determination can be clearly separated from the later steps
of differentiation.

From observations on long-term cultures it is evident that *a specific state of determination can be propagated* over numerous cell generations and is therefore to a certain
extent *stable*. There is considerable evidence that determination is carried by the
individual cell rather than by the organ anlagen as a whole. For that reason the
mechanism for propagation of the determined state is designated as *cell heredity*, but
the carrier of determination is still unknown.

The stability of the determined state is limited. It has been shown that *changes of
determination (transdeterminations)* occur in clones derived from determined cells. They
lead to a new quality of determination which again can be propagated or undergo
further changes. Each cell type gives rise to a limited array of other cell types with a
definite probability. The majority of the observed transdeterminations are *reversible*
and completely stable blastemas which do not revert are rarely obtained. The factors
causing transdetermination presumably reside in the mechanism of cell heredity.
There is accumulating evidence that *cell divisions* are a *necessary prerequisite for transdetermination*. The same seems to be true for a change in temporal differentiation of
epidermal cells in Lepidoptera. For an adult epidermal cell to revert and produce
again a pupal cuticle, it is necessary for the cell to undergo mitosis prior to this
reversed moult (H. PIEPHO, personal communication; KRISHNAKUMARAN et al. 1967).
Since transdetermination is a directed and reversible process which presumably occurs
in groups of contiguous cells simultaneously, we assume that it involves changes of
gene activity rather than changes in the structure of the genetic material.

The main problem for future research seems to be the identification of the carrier
of determination which might give the key to the whole problem of cell differentiation.

References

ASTAUROFF, B.: Studien über die erbliche Veränderung der Halteren bei *Drosophila melanogaster*. Wilhelm Roux. Arch. Entwickl.-Mech. Org. 115, 424—447 (1929).
BALKASCHINA, E.: Ein Fall der Erbhomöosis (Die Genovariation *"aristopedia"*) bei *Drosophila melanogaster*. Wilhelm Roux' Arch. Entwickl.-Mech. Org. 115, 448—463 (1929).
BECKER, H.: Über Röntgenmosaikflecken und Defektmutationen am Auge von *Drosophila*
und die Entwicklungsphysiologie des Auges. Z. indukt. Abstamm.- u. Vererb.-L. 88,
333—373 (1957).
BODENSTEIN, D.: The postembryonic development of *Drosophila*. In: Biology of Drosophila
(Ed.: M. DEMEREC), pp. 275—367. New York: John Wiley and Sons 1950.
—, and A. ABDEL-MALEK: The induction of *aristapedia* by nitrogen mustard in *Drosophila
virilis*. J. exp. Zool. 111, 95—115 (1949).
BRIDGES, C., and T. MORGAN: The third chromosome group of mutant characters of
Drosophila melanogaster. Carnegie Inst. of Washington Publ. No. 327, p. 225 (1923).
EPHRUSSI, B., and G. BEADLE: A technique of transplantation for *Drosophila*. Amer. Natura-
list 70, 218—225 (1936).

GARCIA-BELLIDO, A.: Pattern reconstruction by dissociated imaginal disk cells of *Drosophila melanogaster*. Develop. Biol. **14**, 278—306 (1966a).
— Changes in selective affinity following transdetermination in imaginal disc cells of *Drosophila melanogaster*. Exp. Cell Res. **44**, 382—392 (1966b).
GEHRING, W.: Phenocopies produced by 5-fluoro-uracil. Drosoph. Inf. Serv. **39**, 102 (1964).
— Übertragung und Änderung der Determinationsqualitäten in Antennenscheiben-Kulturen von *Drosophila melanogaster*. J. Embryol. exp. Morph. **15**, 77—111 (1966a).
— Bildung eines vollständigen Mittelbeines mit Sternopleura in der Antennenregion bei der Mutante *Nasobemia* (*Ns*) von *Drosophila melanogaster*. Arch. J. Klaus-Stift. Vererb.-Forsch. **41**, 44—54 (1966b).
— Clonal analysis of determination dynamics in cultures of imaginal disks in *Drosophila melanogaster*. Develop. Biol. **16**, 438—456 (1967).
—, and S. SEIPPEL: Die Imaginalzellen des Clypeo-Labrums und die Bildung des Rüssels von *Drosophila melanogaster*. Rev. Suisse Zool. **74**, 589—596 (1967).
GEIGY, R.: Erzeugung rein imaginaler Defekte durch ultraviolette Eibestrahlung bei *Drosophila melanogaster*. Wilhelm Roux' Arch. Entwickl.-Mech. Org. **125**, 406—447 (1931).
GLOOR, H.: Phänokopie-Versuche mit Äther an *Drosophila*. Rev. Suisse Zool. **54**, 637—712 (1947).
HADORN, E.: Differenzierungsleistungen wiederholt fragmentierter Teilstücke männlicher Genitalscheiben von *Drosophila melanogaster* nach Kultur *in vivo*. Develop. Biol. **7**, 617 to 629 (1963).
— Problems of determination and transdetermination. Brookhaven Symp. Biol. **18**, 148 to 161 (1965).
— Konstanz, Wechsel und Typus der Determination und Differenzierung in Zellen aus männlichen Genitalanlagen von *Drosophila melangoaster* nach Dauerkultur *in vivo*. Develop. Biol. **13**, 424—509 (1966).
— Dynamics of determination. In: Major problems in developmental biology. (Ed.: M. LOCKE), pp. 85—104. New York and London: Academic Press 1967.
—, G. ANDERS u. H. URSPRUNG: Kombinate aus teilweise dissoziierten Imaginalscheiben verschiedener Mutanten und Arten von *Drosophila*. J. exp. Zool. **142**, 159—175 (1959).
—, G. BERTANI u. J. GALLERA: Regulationsfähigkeit und Feldorganisation der männlichen Genital-Imaginalscheibe von *Drosophila melanogaster*. Wilhelm Roux' Arch. Entwickl.-Mech. Org. **144**, 31—70 (1949).
—, u. D. BUCK: Über Entwicklungsleistungen transplantierter Teilstücke von Flügel-Imaginalscheiben von *Drosophila melanogaster*. Rev. Suisse Zool. **69**, 302—310 (1962).
HANNAH-ALAVA, A.: Morphology and chaetotaxy of the legs of *Drosophila melanogaster*. J. Morph. **103**, 281—310 (1958).
HENKE, K., u. H. MAAS: Über sensible Perioden der allgemeinen Körpergliederung von *Drosophila*. Nachr. Akad. Wiss. Göttingen, Math-phys. Kl., **1**, 3—4 (1946).
KRISHNAKUMARAN, A. S. BERRY, H. OBERLANDER, and H. SCHNEIDERMANN: Nucleic acid synthesis during insect development-II. Control of DNA synthesis in the *Cecropia* Silkworm and other Saturniid Moths. J. Insect Physiol. **13**, 1—57 (1967).
LEES, A., and C. WADDINGTON: The development of bristles in normal and some mutant types of *Drosophila melanogaster*. Proc. roy. Soc. B, **131**, 87—110 (1942).
LEWIS, E.: Genetic control and regulation of developmental pathways. In: The role of chromosomes in development (Ed.: M. LOCKE), pp. 231—252. New York and London: Academic Press 1964.
NÖTHIGER, R.: Differenzierungsleistungen in Kombinaten, hergestellt aus Imaginalscheiben verschiedener Arten, Geschlechter und Körpersegmente von *Drosophila*. Wilhelm Roux' Arch. Entwickl.-Mech. Org. **155**, 269—301 (1964).
—, and G. SCHUBIGER: Developmental behaviour of fragments of symmetrical and asymmetrical imaginal discs of *Drosophila melanogaster* (*Diptera*). J. Embryol. exp. Morph. **16**, 355—368 (1966).
PEYER, B., u. E. HADORN: Zum Manifestationsmuster der Mutante "*multiple wing hairs*" (*mwh*) von *Drosophila melanogaster*. Arch. J. Klaus-Stift. Vererb.-Forsch. **40**, 19—26 (1965).

Remensberger, P.: Cytologische und histologische Untersuchungen an Zellstämmen von *Drosophila melanogaster* nach Dauerkultur *in vivo*. Chromosoma (Berl.) 23, 386—417 (1968).

Sang, J., and J. McDonald: Production of phenocopies in *Drosophila* using salts, particularly sodium metaborate. J. Genet. 52, 392—412 (1954).

Schläpfer, T.: Der Einfluß des adulten Wirtsmilieus auf die Entwicklung von larvalen Augenantennen-Imaginalscheiben von *Drosophila melanogaster*. Wilhelm Roux' Arch. Entwickl.-Mech. Org. 154, 378—404 (1963).

Schubiger, G.: Anlageplan, Determinationszustand und Transdeterminationsleistungen der männlichen Vorderbeinscheibe von *Drosophila melanogaster*. Wilhelm Roux'Arch. Entwickl.-Mech. Org. 160, 9—40 (1968).

Stern, C.: The cell lineage of the sternopleura in *Drosophila melanogaster*. Develop. Biol. 7, 365—378 (1963).

— Somatic crossing over and segregation in *Drosophila melanogaster*. Genetics 21, 625—730 (1936).

Sturtevant, A.: The *claret* mutant type of *Drosophila simulans*: A study of chromosome elimination and of cell-lineage. Z. wiss. Zool. 135, 323—356 (1929).

Tobler, H.: Zellspezifische Determination und Beziehung zwischen Proliferation und Transdetermination in Bein- und Flügelprimordien von *Drosophila melanogaster*. J. Embryol. exp. Morph. 16, 609—633 (1966).

Ursprung, H.: Untersuchungen zum Anlagemuster der weiblichen Genitalscheibe von *Drosophila melanogaster* durch UV-Strahlenstich. Rev. Suisse Zool. 64, 303—311 (1957).

— Fragmentierungs- und Bestrahlungsversuche zur Bestimmung von Determinationszustand und Anlageplan der Genitalscheiben von *Drosophila melanogaster*. Roux' Arch. Entwickl.-Mech. Org. 151, 504—558 (1959).

— *In vivo* culture of *Drosophila* imaginal discs. In: Methods in Developmental Biology (Ed.: F. Wilt and N. Wessells) pp. 485—492. New York: Thomas Crowell Co. 1967.

—, u. E. Hadorn: Weitere Untersuchungen über Musterbildung in Kombinaten aus teilweise dissoziierten Flügel-Imaginalscheiben von *Drosophila melanogaster*. Develop. Biol. 4, 40—66 (1962).

Waddington, C.: Principles of embryology, p. 349. London: Allen & Unwin 1956.

—, and M. Perry: The ultra-structure of the developing eye of *Drosophila*. Proc. roy. Soc. B, 153, 155—178 (1960).

Wildermuth, H.: Differenzierungsleistungen, Mustergliederung und Transdeterminationsmechanismen in hetero- und homoplastischen Transplantaten der Rüsselprimordien von *Drosophila*. Wilhelm Roux, Arch. Entwickl.-Mech. Org. 160, 41—75 (1968).

— Autoradiographische Untersuchungen zum Vermehrungsmuster der Zellen in proliferierenden Rüsselprimordien von *Drosophila melanogaster*. (Develop. Biol. 18, 1—13) (1968).

—, u. E. Hadorn: Differenzierungsleistungen der Labial-Imaginalscheibe von *Drosophila melanogaster*. Rev. Suisse Zool. 72, 686—694 (1965).